THE
HAPPINESS MYTH

THE HAPPINESS MYTH

WHY WHAT WE THINK IS RIGHT IS WRONG

A History of What Really Makes Us Happy

JENNIFER MICHAEL HECHT

HarperSanFrancisco
A Division of HarperCollins*Publishers*

HarperCollins books may be purchased for educational, business, or sales promotional use. For information please write: Special Markets Department, HarperCollins Publishers, 10 East 53rd Street, New York, NY 10022.

HarperCollins Web site: http://www.harpercollins.com
HarperCollins®, ■®, and HarperSanFrancisco™ are
trademarks of HarperCollins Publishers.

FIRST EDITION
Designed by Joseph Rutt

Library of Congress Cataloging-in-Publication Data is available.
ISBN: 978–0–06–081397–0
ISBN 10: 0–06–081397–0

07 08 09 10 11 RRD(H) 10 9 8 7 6 5 4 3 2 1

To my husband, John;
our son, Max;
and our daughter, Jessie Leo.
In happiness I could not have predicted.

Contents

Get Happy
Myths of the Modern Mind

In the four hundred years from the beginning of the sixteenth century to the end of the nineteenth, women wore corsets to shape their figures. At first the ideal was to have a flat-front, cylindrical shape; later the ideal was a gentle hourglass, big at the breasts and hips, and smoothly slanting in toward a small waist. Corsets also helped support the heavy dresses that women wore in this period, which often included bustles at the hips and rear end, scaffolding out a hugely exaggerated frame. If the fabric creased anywhere, it hurt, causing too much pressure in one place, so corset makers put in structural elements to keep the pressure distributed across the whole torso. These were *stays*, made from steel, whalebone, or ivory. Corsets were always snug enough to be supportive, but in the nineteenth century the fashion of *tight-lacing* began, giving women a shape that looked like an actual hourglass. With tight-lacing beginning in puberty, a woman's ribs grew compressed, forcing her lungs, liver, and heart to be displaced and squeezed, and her intestines could barely move enough to function. The pressure on women's lungs caused a lot more fainting than might otherwise have occurred, and there were occasions of serious organ damage—even of prolapsed uterus, where the pressure forces the uterus down into the vagina.

Today we wear soft cotton shirts and pants, and women get support for their breasts from other supple materials designed for comfort. But the same woman who pities the corseted girls of the past may very well go to the gym five times a week to do a hundred sit-ups on a Swiss ball and an hour of aerobics. Most likely she is exacting about her diet, eats foods that have been stripped of various life-sustaining elements, and thus keeps her body fat to a minimum. The whalebone ribbing is gone from just beneath her dress, but now her own ribs show through the cloth! A modern woman may wear a breast-enhancing bra, or even have surgery to install what amounts to bulbous bustles just beneath the skin of her bosom. She may damage her body in her exertions—knees and wrists seem particularly vulnerable—and she may lose control of the dieting and find herself unable to eat normal meals and maintain health, particularly reproductive function. Also note that though it was hard to fail at wearing a corset, many women today fail at their body goals, spend massive amounts of money on half-used gym memberships and diet programs, and, worst of all, feel tortured by their failure. Indeed, other people will judge a woman with a slim body to be happier than the woman "trapped" in her fat.

So who is crazier, the culture that had women bind their bodies, or us? Even when both of these practices are carried out in moderation, does it make more sense for us to expect women to come in various sizes but to squeeze them all into the same shape, from the outside, or does it make more sense for the culture to accept various shapes but idealize the lack of body fat, forcing people to exert all sorts of internal pressures on themselves? We have a commonly held belief today that corsets were cruelly uncomfortable and bizarre, but our own practices do not suffer a comparison well. History may be read as a source of anthropological insights, but we can also turn the light of those insights on what we do today. It is not easy to find the relevant connections, but if we come to comprehend an Old World point of view as ordinary—the absurd becomes normal—the effect on how we view our modern world can be profound: the normal becomes absurd. Often, the modern point of view reveals itself to consist of bossy, shaming, controlling nonsense. This realization alone can give us a bit of freedom from the mental corsets of our day. Most of the strictures we live under are just cultural stories, no more inherently true than the cultural stories of any other period in history.

Our times are not exactly like any other time in history. We are very special. But not as special as we think. The fact that you can send a probe to Mars doesn't mean you are better at living a normal life. It is true that modern experimental science can turn up new kinds of information, but brain scans and scientifically constructed questionnaires answer only the questions that you ask. Happiness is a subjective matter. How superior, then, are we to those of earlier centuries who analyzed happiness using observation and contemplation? Historically, people have studied happiness from their own point of view. That may seem subjective, but when you yourself judge the happiness of fifty people you have known, it is always the same person playing judge. That is a big benefit. When, in the style of modern experimental science, we ask the fifty people to report their own experiences, they may each have such varied expectations that their reports are meaningless. Such studies have shown that a lottery winner and an accident victim left paraplegic both return to their genetic happiness "set points" a year after the event. Historically, we have always known of "happy types," but the modern idea of set points may go too far.

Perhaps you can look at your life and say that you were reasonably happy in most periods. But ask yourself whether you were happier when things were going your way than when you suffered a major setback. We may thank science for reminding us that good and bad luck do not change things as much as we fantasize and worry that they will. But the old intuitive pattern—good stuff is better—holds up well. Our present-day affection for the idea of happiness "set points" also seems a bit disrespectful to the countless people who have left records saying that events or society had stolen much of the happiness they should have had. The evidence for a "set point" is a study of identical twins separated at birth, some of whom seemed to have very similar mood ranges. The evidence against it comes from millions of attestations over thousands of years. Our science of happiness can be fascinating, but it is not definitive. Our forebears had other ways of judging what makes people happy, and future generations will have others still. We may think we are much more sophisticated than people in the past, but the happy truth is that we can learn a lot from history.

Our era believes that happiness requires a particular relationship to physical exercise, legal and illegal drugs, food, public decorum, sexual normality, and material and media culture. It has no more claim to truth on these matters than any other. People argue about the details but accept the basic modern assumptions as true, based on science and common sense. This book seeks to prove that the basic modern assumptions about how to be happy are nonsense and that we do not have to listen to people argue about the details. That is a big claim, I know. But I am convinced both that it is accurate and that it is a wonderfully therapeutic idea if you can fully take it in. All disciplines, from physics to comedy, may be said to progress, but all disciplines do not equally generate useful ideas that must be understood by the next generation if the game is to be advanced. Once the whole idea of what we are doing changes, I no longer need to know the details of the old obsession. The circumference of a Caucasian head, the number of changes in a bluegrass song—anthropology and popular music go on, to varying degrees, without needing or wanting this information. We think of our version of a happy life as more like physics than like pop songs; we expect the people of the next century, say, to agree with our basic tenets—for instance, that broccoli is good for a happy life and that opium is bad—but they will not. Our rules for living are more like the history of pop songs. They make their weird sense only to the people of each given time period. They aren't *true*. This book shows you how past myths functioned, and likewise how our myths of today function, and thus lets you out of the trap of thinking you have to pay heed to any of them. This process of examining myths is also good for sharpening our ability to see truths other than our own. We get to lose some respect for our own dogma, and to gain some respect for the dogma of other historical periods. This is a necessary and sufficient characteristic for understanding other historical eras—that is, for doing good history.

If you are looking for advice on happiness, the usual thing to do would be to find a book written by a psychologist. Most psychologists believe that people get locked into certain behaviors and feelings because of the way the world treated them when they were young. Let's say you are anxious and demanding. Typical psychological advice would be that a whole group of people feel and act the way you do and that many individuals of this group lived in the same kind of world as yours when they were young, maybe with unreliable parents. If the advice is very insightful, or if you just get lucky, you might have a big realization.

We all believe that we have to act the way we do in order for things not to come crashing down around us, but how often do we check? A big realization can show us that our guesses about the world are skewed — in this example, skewed toward thinking that other people will let us down. The realization might allow us to stop monitoring the shopping list or the office politics and just see what happens if we let everyone else do things their way. Likely as not, people will come through — on their own terms, but substantially. We all know it is exceedingly difficult to talk people out of their personality traits. But when such efforts are success-ful, it is because a person becomes aware of a viable alternative to their way of seeing the world. Instead of looking at the world through a nar-row tube, now they can look at the world through two narrow tubes! Later, with some imagination, they may notice that there must be about six billion narrow tubes, and that big-picture reality will be available to them not by calling all the other assessments of the universe crazy or stupid, but rather by trying to accept them all as informative, and to use them all to build a patchwork vision of the whole.

Marvelously, history can work the same way. A lot of people read his-tory so that they can get a glance at another way of seeing the world, another way of being a human being. History, too, has its explanatory methods. On the matter of history's corset bones versus today's skin and bones, we may note that in eras of intense family hierarchy, authoritari-an governments, and dogma as an intellectual ideal, it seemed right for clothing, also, to squeeze and support you from the outside. In an era of individual rights and responsibilities, clothing is soft and accommodat-ing, but people are told to do a lot of intense internal squeezing and controlling. It is a modern myth that a person with a lithe, toned body is better than other people, happier and healthier. We think of beliefs such as this as modern, and therefore more scientific than the claims of the past. But today, too, the cultural code of how to live and how to feel about things is just an unscientific web of symbolic cultural fantasies.

There are many ways that the past seems controlled by its crazy myths. Think of the special "blue blood" of the aristocracy — a whole society, over a thousand years, based on the idea of special blood! Or consider how the actual lives of people through history have been domi-nated by the cultural myth of the sanctity of virginity. Consider the con-viction that making money is vulgar. We think of ourselves as living more reasonable, happier lives because we are free from such myths and able to base our behavior on common sense and empirical science. But that isn't true: we are swamped with cultural messages that are as subjective as those of any other moment in history. And when our culture determines our approach to happiness in a way that seems, in light of history, particularly limiting, it is right to take a good look at the situation.

Consider, for instance, that the difference between a society that encourages sad people to sip opium drops and one that encourages them to take Zoloft is not "the advance of science." Rather, the difference is that nowadays everyone drives cars and handles money. Clearheadedness is more important now than it ever was. We devalue euphoria in our drugs—describing bliss as an unwanted side effect, like hair loss—because we value productivity. When it comes to the care of the body, our scientists search for longevity and not for euphoria. If experimenting on chocolate cake, our scientists will study how quickly it kills you, not whether the pleasure is worth it. These days all the bliss drugs are taken under the cover of another action—killing pain or inducing sleep—or outside the purview of legal medicine. This cultural bias toward productivity and longevity is so strong that it makes our euphoria decisions for us! The huge assumptions about happiness that guide our actions are based on myths, fantasies, cultural hypnotic trances. We have to snap ourselves out of them. Our "trances of value" keep us from things that might make us happy; indeed, they make us feel conflicted about the way we negotiate the choices that are left to us.

Historical perspective can so alter our assumptions as to make genuine happiness more available—sometimes in the flash of a realization. For instance, I'll bet you think that the happiness you get as a result of drugs is less valid than sober happiness. I'll bet that when you think, while in a period of drug-assisted happiness, "Gee, I've been happy a lot lately," you hesitate to compare this time favorably with other, drug-free times in your life. Yet not all cultures tell people that drugs create fake happiness. How could drugged happiness not be as real as any other kind of happiness? The idea that drugs create fake happiness is a prejudice, an assumption about value. Historical study of the emperor and philosopher Marcus Aurelius's opium habit makes me wonder why we are so convinced that drug-related happiness is not an authentic happiness. On some subjects, I fear I will sound like an apologist for indulgence, but I do not think it is morally justifiable for an able-bodied man or woman to devote huge amounts of time and energy to worrying about things that do not really matter much. Don't you know people who are very proud, for instance, of their "healthy" diet, and other people who are very ashamed of their "unhealthy" diet? Shouldn't these people be proud and ashamed of something of more substance? I am embarrassed by how pathetic some of our priorities will look to the future. I think it will be clear to future historians that these mythic obsessions with the body are responsible not only for the bony fashion model, but also for the extra-large average person, for whom eating becomes an expression of rejecting these shaming forces of control. And it will be clear that our myths about drugs are responsible for a lot of unhappy drug taking and a lot of unhappy drug abstaining. I believe that a moral imperative to be of use begins with a moral imperative to get one's mind right, to be able to see nonsensical cultural assumptions—trances of value—for what they are, to develop oneself as a truth detector. There is no reason to think that we can each individually do this crucial work on our own, without scholarship, and without carefully sketching out just what it is we think we mean—and what it is we want.

There is much in our daily lives that is hidden from us because we are too close to it. We need historical comparison in order to make our own rituals visible. For instance, historically, whole towns have celebrated joyous and mournful holidays together, with everyone in town aware of the same poignant mythic story: of an ancient god kidnapping another's virgin daughter, or of a young pregnant woman on a donkey, searching for an indoor place to birth her extraordinary son. Where is today's mythic image of a girl in trouble? In the news. We follow the news, constructed out of the infinity of human events into a genre all its own that is a running loop of wounds, disappearances, deaths, and rescue. We oddly call this loop of lost girls and other iconic troubles *news*—the same word we use to indicate the details of politics, economics, culture, and science that affect us all. It may be silly to call it news, but the sensationalist news has its own function as a story of grief and occasional recovery or rebirth, a story that all of us know, that all of us can celebrate, and that all of us can lament. I believe that this story works to bind the community on the level of shared knowledge and conversation, and that it can much more powerfully affect our happiness if we actually join the parade, demonstration, or vigil. Furthermore, when we do go to community celebrations, we get the most out of them when we know about how they worked in the past.

We are not individuals, not really. We are each a node in a mesh of relationships. Throughout history, a lot of group behavior was mandatory: we were all locked into our roles in our extended families, our religions, and our towns. For much of the twentieth century, our culture was marked by participation in voluntary associations, from the Elks club, to trade-union halls, to the numerous women's auxiliary clubs, to the Boy Scouts and Brownies. Today we mostly participate in one-day commitments. For many of us, that is what works nowadays. Many of us do not even do that much. But we miss out on a lot of what makes people happy if we do not investigate joyful public celebrations and public demonstrations of anger or grief. And we are missing a lot of what makes people happy if we fail to recognize that in attending an annual public event, we are keeping that event, and its community, alive. We embody something larger than our own lives, and when we take part in public expressions of emotion, we are doing something valuable for our human hearts. This may seem an odd assertion, but I think that, once we have a sense of what the ancient and medieval worlds actually did at their celebrations, it becomes obvious that we need some of that now. We certainly need to rebut those cynics who think of all such actions as simple or shallow. Indeed, those actions are a primary foundation for the building of political morality and agency, and for personal solace.

Like celebration, there are other much-maligned aspects of our culture that seem more humanist when seen in light of historical behavior. Shopping, watching television shows, and following sports teams are all ways that people today take part in an otherwise painfully heterodox culture. That is to say, we do not share religion, politics, or ethnicity with the people we work with. We are in need of a common culture that allows for some personal expression, but one that is basically safe. How we accept or reject that culture determines a lot about how we feel about the world, how happy we are. It is a modern myth that money cannot make you happy. We all say that it can't, but, given one wish, a lot of us would go for cash. We certainly opt for money over many other pleasures in structuring our real lives. Part of the reason is that what you can buy with money today you used to be able to get for free—social contact and play that can fit neatly into your life. Shopping, television shows, and sports are not deep, but neither were the common social contact and play that kept people happy in the past.

These four issues—drugs, money, bodies, celebration—seem to me the largest topics of happiness that are made visible through historical study, and they are at the center of this book. What about love, faith, and art? They are not made visible through historical study. Love, faith, and art seem to me safe from the chuckles of history: if you have passion in love, faith, or art today, you are likely to take seriously an expression of love, faith, or art from two hundred years ago, or five hundred, or two thousand. We understand a toy carved by a frontier settler for his infant son; we understand a cathedral; we understand a gazelle painted on the wall of a cave. A doctor's advice, by contrast, does not hold up even from decade to decade. Nothing is more important for your happiness than whether you are fully interacting with the timeless realm of things. Much can be done to enhance that interaction: be more open to change and more forgiving of faults; find time to do some painting; go to church or temple, or go to the ocean; go to the planetarium and see an intergalactic show; find some way to help someone; or be an inspiration. This book trusts you to know as much about the timeless sources of happiness as anyone else. But very likely you have a lot of trouble doing all of this. It is as if you know the beautiful ballroom dance, but you are limping. Well, the limp is curable through history. Drugs, bodies, money, and celebration are the themes of the four central chapters in this book. They may seem an odd collection of subjects. Maybe they are. They all overlap so much that I certainly could have divided my discussion of them along other lines. Still, these do very nicely. Happiness is feeling good. Drugs, bodies, celebration, and money are worldly, mundane techniques for making ourselves feel good. They are badly controlled by historical whims, and we need to rise above today's version of mythic prattle about them and still take advantage of what they have to offer. I call my method "happiness by historical perspective."

The book opens a little differently. The first section is a summation of happiness wisdom through all of time. It highlights the ways in which our society has taken certain age-old advice and blown it far out of proportion while giving short shrift to other advice. In this sense, this is another kind of happiness by historical perspective. But this section is also an opportunity to collect the greatest pieces of advice for individual happiness from all time and sort them into four memorable directives.

Happiness may seem like a populist goal, as if serious people should consider less emotional matters. But since we judge everything through our minds and moods, what could be more essential as a philosophical starting place? Ignore the reputation of "self-help," and let us concede that happiness is important to us. As I see it, there is a moral imperative to be open-minded, investigative, and in pursuit of happiness. People have argued that we moderns are self-indulgently expectant of our own happiness, compared with the great mass of historical humanity who lived without such lofty goals. But I do not think this is true. Indeed, I believe that other historical eras have given more value to happiness in the form of euphoria, leisure, play, celebration, and indulgences. Other eras have also offered more choices of behaviors with no end other than the happiness of a given individual or crowd. Yes, there was also more hardship in the past and more hierarchy, and I do not want to go back. But I do think we have a lot to learn from other historical moments.

What is happiness? Obviously, we are dealing with an issue that is a moving target. We could say that everything we welcome is an agent of our happiness, including misery—and we will have lost our subject. We know what we mean when we say the word in context, but it has a cluster of meanings that must be sorted out. There are three distinct kinds of happiness, and though they are not unrelated, they are not usually in harmony with each other.

A *Good Day*   A good day can be filled with many mild pleasures, repeatable and forgettable, and some rewarding efforts.

Euphoria   Euphoria is intense, lasts powerfully in memory, and often involves some risk or vulnerability.

A *Happy Life*   A happy life requires a lot of difficult work (studying, striving, nurturing, maintaining, negotiating, mourning, and birthing), sometimes seriously cutting into time for a good day or for euphoria.

Anything we do may facilitate one kind of happiness and inhibit another. Researchers have plaintively wondered how people can report that they are so happy watching television when by other tests we find that they are only semiconscious when doing so. Yet every one of these people will tell you they prefer life to death; they want to be here. The answer is simple: *The three kinds of happiness are not only very different; they are often at odds.* They can be united in one experience, but more often than not a euphoric experience is also a painful or difficult one; a good day includes more playing than would add up to a happy life if you did it all the time; and arranging for a happy life is effortful and often unpleasant. We get confused when we forget that these three kinds of happiness can rarely be served at once. If you live every day as if it is your last, you miss out on all the happiness connected with effort and building, and will be struck by all the trouble that euphoria entails. If you attend only to one kind of happiness and find yourself generally unhappy, you should look to the other kinds.

Modern expert advice, with its obsession with longevity and productivity, is hopelessly devoted to "happy life" happiness. When we reject expert advice, it is in deference to the other kinds of happiness. A lot of what looks like a lack of willpower is, from another perspective, a series of positive choices in favor of one of the other kinds of happiness. Again, I risk seeming to be an apologist for laziness or bad behavior, but I think the risk is worth the danger. This is happiness we are talking about—too important a subject to fudge at the edges for the sake of propriety. The content of happiness advice in our culture features some fascinating schisms regarding how real people—scholars, jocks, jerks, and grandmothers—actually behave. We all act like we fall short of various ideals to various degrees. Isn't that strange? Compare how you feel about seeing a policeman on the street as you walk through town with how you feel when you are driving and you spot a police car; most people walking down the street are not breaking any laws, but many people do exceed the speed limit while driving, and do make the occasional illegal turn. When you drive, you accept the cultural assignment to feel a bit guilty when you see the police. Regarding our own happiness, we have accepted the cultural assignment to feel a bit guilty.

The commonly expected story is that, though we all accept the values of happiness advice, we fall short because of failures in self-control. A common idea from Augustine to Freud, and one still generally held today, is that within us are several people fighting each other: a party animal, a drill sergeant, and a heart-of-gold observer. It is a fruitful metaphor, but of course we are not several people, but one person making decisions about life's many choices. The information about those choices changes by the moment. On Monday morning you decide to renounce eating cake until you are as thin as you were in youth. On Monday morning this decision serves long-term happiness, and without much cost, because you are not in the mood for cake anyway, and you recognize that the loss of a little pleasure in any given day can easily be tolerated. But on Thursday night you have different, equally valid information: now you very much desire some cake; indeed, it seems that you could give yourself a big rush of good-day happiness, possibly even bordering on the euphoric. The logic of the decision changes—which is why the same sane person makes two different happiness decisions, one on Monday morning and another on Thursday night. One of these isn't the Good Adult and the other the Bad Kid. That is culture talking. Each of us strikes a balance between good-day happiness and happy-life happiness—and occasional euphoria. Only the *you* of any given moment knows the relative values of each kind of happiness for that given moment. Making decisions in opposition to expert advice is not cheating.

Euphoria is the spice of life: you don't need much, but most of us really need some. There are people who are religious their whole lives, carried through by a personal experience of the holy that lasted twenty minutes in 1978, when they had a vision while on a hike at church camp. There are people who base their philosophical understanding of the universe on three ten-hour hallucinogenic-drug trips they took in college and the subsequent reading of an essay on Kant. Intense experiences are full of information and can last a lifetime. The memory of euphoric sex has carried many couples across long periods of dearth. One day of a blissful community festival can forge allegiances to an idea or a place that can go on to animate a lifetime. The emotion of ecstasy is shockingly potent stuff. It is a good thing the effect lasts so long, because ecstasy is not easy to obtain. There are only four kinds of ecstasy in our common lexicon: drug induced, sexual, spiritual, and bacchanalian. They are all hard to get, tiring to keep, and rare to have repeated, and they may be followed by some sadness, or at least dry mouth and a headache. Certain subsets of the population also try to achieve extreme bliss through physical effort and danger (warfare and dangerous sports) or through the heights of artistic experience. But mostly, the term *ecstasy* is associated with intoxication, orgasm, mystical elation, or ancient festal frenzy. In any form, ecstasy is either hard to invoke or comes with a price of pain and real danger. There are domestic heights of joy—weddings, for example—but note that they are self-limiting by nature. We can create more joy if we try: in the mid–twentieth century, the psychologist Abraham Maslow offered ideas about how we could cultivate more "peak" moments. But there is a point where the effort becomes like trying to make rarity more common. This is okay, though, because you need only a few euphoric events to flavor a whole life with happiness.

This is what I mean by a careful sketching of ideas. I think we all make these kinds of analyses on the fly, every day, as we wake up and decide what to do with ourselves. I also think that having some theoretical structure can help people make better decisions, perhaps new decisions, with more understanding and fewer regrets. If you wanted to peer into the universe or the atom, you might find yourself without the right lab equipment; but if you want to be a scientist of happiness, you have already been assigned a testing subject—yourself. But you *do* have to actually carry out experiments; that is, just thinking about them will not suffice. We today have all sorts of reasons for not going to the parade or the spa, but if you want to know what kind of effect these places have on people, you have to go. Behaviors that have stood the test of time as conducive to happiness have got to be looked at carefully. Although a lot of them are available to us, we do not understand them as happiness vehicles, and thus we either avoid them as a waste of time or we engage in them, but without making ourselves aware of their potential to make us happy. Once you know that caffeine was powerfully associated with happiness in the days, and centuries, before our devotion to productivity, you will find yourself experiencing coffee differently.

We need to pay careful attention to our modern, unhelpful myths so that we can make better choices. The way we do happiness is too constrained by faulty assumptions. These are not matters we can choose to either deal with or ignore. They knock us about, on a daily basis, and all through our notoriously long lives. It would seem worthwhile to try to get a hold on them. Historical examples help. Sometimes the lesson is to go out and change our behavior, and sometimes a remarkably different experience of the same behavior becomes possible with the simple addition of some big-picture knowledge. Men once wore unwieldy top hats, suggestive of the tall forehead of a mature man; but now it is youth that is correlated to ability, so men have hair sewn into their naked scalps. Yet the worst thing we do to ourselves is not about giant hats, tight corsets, bad diets, or cosmetic surgery; the worst thing we do to ourselves is to let ourselves get trapped in thoughts and behaviors for decades, for a lifetime, repeating and repeating. What a paralyzing potion culture can be! The antidote is history.

Wisdom

"What good can we suppose it did Varro and Aristotle to know so many things? Did it exempt them from human discomforts? Were they freed from the accidents that oppress a porter? Did they derive from logic some consolation for the gout?" I wrote the book you are now reading because of these sentences, written by Montaigne in 1576.[1] The passage continues:

> For knowing how this humor lodges in the joints, did they feel it less? Were they reconciled to death for knowing that some nations rejoice in it, and with cuckoldry for knowing that wives are held in common in some region? On the contrary, though they held the first rank in knowledge, one among the Romans, the other among the Greeks, and in the period when knowledge flourished most, we have not for all that heard that they had any particular excellence in their lives; in fact the Greek has a hard time to clear himself of some notable spots in his.[2]

Montaigne asks if any happiness can be expected from learning. "Have they found that sensual pleasure and health are more savory to him who knows astrology and grammar?" And he continues: "I have seen in my time a hundred artisans, a hundred plowmen wiser and happier than rectors of the university."[3] It is not only rote learning that he disparages, but even the wisdom that is supposed to come from knowledge. He claims that ignorant men surpass learned men in every virtue of action and conduct. Considering these lines of Montaigne's, I wondered why I and other professors so confidently insist to our students that philosophy, wisdom literature, and even general knowledge will make them happier. And what about all the very smart, educated, miserable people I know?

Koheleth, the author of the book of Ecclesiastes, also shocks the modern reader with his lack of doting respect for knowledge and wisdom. The second section of Ecclesiastes is called "Wisdom Is Meaningless," and in it Koheleth tells us that in his own life, he devoted himself "to explore by wisdom" everything "under the sun" and this is what he learned from these paths of wisdom: "All of them are meaningless, a chasing after the wind. What is twisted cannot be straightened; what is lacking cannot be counted ... with much wisdom comes much sorrow; the more knowledge, the more grief." Koheleth acknowledged that "wisdom excels folly as light excels darkness" but that, in the long run, light and darkness come to the same end. "The wise man's eyes are in his head, but the fool walks in darkness. And yet I know that one fate befalls them both ... the wise man and the fool alike die!" His exhaustion had a giddy quality to it. "Be not righteous over much," cautioned Koheleth, "neither make thyself over wise: why should thou die before thy time?" (Eccles. 7:16). Koheleth and Montaigne were scholars, and they joked about knowledge as one jokes about one's beloved spouse. Still, a lot of the wisest men and women try to warn us that knowledge is not always an aid to happiness, and that even insight and wisdom can be useless against a dark mood.

Half the rank and file of humanity is too sensible to bear with the optimism, certainty, and narcissism of the self-help guru. The other half is trying for relaxation and doesn't mind submission as a fast way to achieve it. Philosophers and authors of wisdom literature are easily differentiated from self-help gurus in that they are too wise to offer much optimism or certainty, and they thus risk losing clients. The guru stays cheerful; know him by his grin. And the narcissism! Narcissism is such a tough problem that the foundational spiritual idea of the entire eastern hemisphere of our planet is that, should you ever overcome your narcissism, so great would be the event that you would pop out of existence in an ecstasy of happiness. That is what nirvana is: the final realization that the self is not what it seems to be. Not only should we not coddle the self and coo at it, nursing its little embarrassments and beaming when it gets its way; we should not even tolerate its existence. We must set upon it, dissolve it, pull it into little strings. Nirvana sounds like an optimistic concept, but when you get up close—that is, when you study the masters—there is a distinct flavor of *sour charm.* This is a term I needed to invent for this book; it means a kind of cheerful world-weariness—not the sad and tender feeling of bittersweet, but a wry, disappointed geniality. The twentieth-century Tibetan Buddhist teacher Chögyam Trungpa wrote: "The attainment of enlightenment from ego's point of view is extreme death, the death of the self, the death of me and mine, the death of the watcher. It is the ultimate and final disappointment." Sour charm.

Most opinion and knowledge is not going to make you happy, and certainly very little of it is both significant and true. Montaigne says we believe whatever they believed in the place where we grew up. That's it. There are no real cultural opinions, just local assumptions. He says we cling to these as to a rock in a storm and wonders just what it is we think will happen if we let go. How much can you despise someone for believing something you would believe if only you had been raised there? Yet if we visit a bunch of these rocks and start to notice something in common among them, the common theme seems worth noting. Despite their many opinions about what we should do with our lives once we get our happiness under control, the philosophers, the wisdom writers, and the self-help leaders all say the same thing about what we should do to get to happiness. That is why self-help leaders can indeed help many smart people, and why even the wisest of us might find an insight among the sugary encouragements and tough love. There are four doctrines found in all happiness theory from wisdom literature, philosophy, psychology, and self-help. They are:

Know yourself.
Control your desires.
Take what's yours.
Remember death.

This is the core, classic wisdom about happiness. It is very difficult to follow any of these doctrines, and when you do make progress in any one of them, that very progress brings new problems. For instance: coming to know yourself can make you vulnerable, controlling your desires can make you passionless, taking what's yours gives you tremendous responsibility, and remembering death can make you too detached to be of full use to yourself and the people around you. That is why it is good to find a guide; that way, you do not get stranded in these classic errors. Despite their risks, these are the magic formulas, and, to an important degree, they work. This section will speak to each of these in turn: how they work, how to work them, and why it is not appropriate to be unbrokenly optimistic about any of it.

Great happiness philosophers all address the four doctrines (self-knowledge, self-control, self-realization, and awareness of death), but they diverge on what else to do with life. There is a busy crowd of suggestions. Ecclesiastes suggests we be devoted to our spouse and to some project of our own making; Plato suggests we follow the pursuit of truth and tend to the health of the community; Epicurus suggests friendship, sex, food, and wine; self-help suggests goals of relaxation and prosperity; much psychology suggests productive labor and a reproductive family; religious literature suggests that people find happiness in devotion to God; much of women's literature suggests that people pursue happiness through nurturing children and others; men's literature suggests finding happiness through competence and competition; children's literature suggests that happiness is to be found in imagining things. Koheleth, Epicurus, and Spinoza tell us to have a good time—that it is part of our job to enjoy pleasure and to create joy. "Eat, drink, and be merry," wrote Koheleth. Spinoza asks the great heroes of self-denial, "Why is it more seemly to extinguish hunger and thirst than to drive away melancholy?" The Buddha, Epicurus, Augustine, and Petrarch tell us it's okay to shun politics and even live outside normal human society, yet we get the opposite message from Aristotle, Cicero, Thomas Jefferson, Karl Marx, Susan B. Anthony, and Martin Luther King Jr., who all tell us to try to help our fellows, even if it is dangerous. Bertrand Russell said that he found the happiness of parenthood greater than any other he had experienced.[4] Artists speak about the ecstasy of creation, the passion hidden in the image of a painter "lost in his work." Modern "positive psychology" emphasizes dedication to labor that is satisfying in the moment. Mihaly Csikszentmihalyi's *Flow: The Psychology of Optimal Experience* of 1990 argued persuasively that people are happiest when engaged in tasks that they get "lost" in, where time just flows: the talented cellist, creating her own bliss. These high-level questions about what we should do with life come to no agreement; all the philosophers have tastes of their own. Yet—what luck—philosophers generally agree on our four core ideas that underlie all that variety, the ideas you have to master in order to do anything else.

1

Know Yourself

Know yourself. This is the key to all philosophy, the center of all wisdom, the one thing that decides if you are the actor in a tragedy or
a comedy. This chapter points out three major interpretations of this singular injunction. The first is the Socratic, and it has to do with knowing
what you believe. The second is Freudian and has to do with knowing who you are. The third is lonely and has to do with training yourself
to take your intellect as your own companion.

In the *Apology*, Plato has Socrates explain that the only happiness is
figuring out what real virtue is, and enacting it. People who behave
badly may seem happy, but they are not, no matter how rich they get,
and people who act with virtue are certain to come into happiness and,
very likely, come into money as well. As he put it: "I do nothing but go
about persuading you all, old and young alike, not to take thought for
your persons or properties, but first and chiefly to care about the greatest
improvement of the soul. I tell you that virtue is not given by money,
but that from virtue comes money and every other good of man, public
as well as private." Coming to know yourself and re-creating how you
experience the world is a more efficient way to get comfortable than
directly altering the world.

An angry person on the subway scowls and pushes, other people
scowl and push in response, and quarrels ensue; a smiling person offers
seats, takes inconveniences with patience, offers to share cabs, and has

merry encounters. The angry person has no idea how much his or her anger colors the way other people act. A sunny disposition is no guarantee they won't steal your wallet, but some of what we don't really know about ourselves gets bounced back from the world and radically conditions how we see things. The Socratic claim that the unexamined life is not worth living is so commonplace that we forget how harsh it is. Vicious even. Think of all the good, sweet fools you know! Isn't it possible to be a decent, gentle, productive person without a jot of philosophy or self-examination? The Socratic answer is resolutely no; the examination of oneself and one's manner of living is the only good life and only cause of happiness. The happiness thus achieved cannot be stolen away by any means. Given the pitiless vagaries of life, the internal nature of philosophical happiness is one of its big selling points.

Socrates insisted that we ask ourselves how we know what we believe. You like democracy, monogamy, American food, sleeping at night, children raised in families, longevity as a life-defining goal. You like a woman of five foot ten to weigh about a hundred forty pounds. Set a goal of convincing yourself of something you oppose. Pick a hot-button subject, and a reward for yourself if you can shake your own faith in your convictions. I have strong political convictions, but I'm not rallying for them right now. I'm suggesting you pull a Socratic trick on yourself and ask yourself all the questions you usually avoid thinking about. If the thought is unbearable, it tells us something about the way we believe, and think, and live. We live in little cognitive comas. Or rather, we cavort in cognitive fields surrounded by electric fences: we all think we are free to go where we wish, but we are struck by a lot of pain when we try to think past our boundaries. Politics are real, but the odds are that if you had been raised in a different U.S. state (let alone China!), you would be not the Democrat or Republican that you are now, but instead a Republican or Democrat. Even though those people make your blood boil. Odds are odds. If you want to know yourself, you are going to have to rough yourself up a little. Socrates and Plato both held that this kind of ruthless thinking makes you happy in the process. When Plato does imagine an arrival, a coming to the most profound knowledge, it is blissful. But most of the time this is all about happiness as a process, as an effort.

Note that philosophy is unlikely to be effective if you just read it. Socrates so believed that philosophy required conversation with others that he did not write any books, and when Plato recorded Socratic thought, he did so in the form of dialog. Many of the great Socratic dialogs took place at social events; the title of Plato's *Symposium* means "the drinking party," and that is where it is set. In a sense that book is one of the most idealistic visions ever crafted, and it took place amid food, copious wine, and modest revelry. How do you *do* philosophy? Discuss it with others, write about it, get locked away with it. The last is the least effective, but it cannot be entirely rejected, because it does work for some people, some of the time. The essence of the philosophical experience, the active verb of *doing* philosophy, is unlearning what you think you know. And it is much easier to find out what your deep assumptions are if there is someone else there to help you discover them. Alone, your best bet is to try to write what you think, and proceed with scrupulous honesty, imagining your own most skeptical self as the reader. Think of the biblical story where Jacob wrestles all night with an angel and the angel wounds him, and changes his name from Jacob (*"who grasps"*) to Israel (*"who prevails"*). Renamed, he can finally ask for his brother's pardon for stealing his birthright, and thus be reunited with him. When you come to something you can't explain, do not gloss over it; stay with it, wrestle it. Confusion is your quarry. Rejoice when you find it, bear with the pain it inflicts, and don't let it go until it gives you a new name. By the way, later, the sun, that symbol of true wisdom, heals Jacob's injury.

Ancient ideas of knowing yourself were about coming to be a better person. The process was psychological, but more in the realm of conditioning one's mind than in finding out why the mind does what it does. Marcus Aurelius said, "Cast away opinion and you are saved. Who then hinders you from casting it away?"[1] Can we really control our emotions by decision? The best of the ancient writers, including Aurelius, acknowledged that we could not do it, and with a smile and a shrug provided exercises for teaching ourselves to improve what self-control we have. That's what religion and graceful-life philosophies are doing with their rituals and their meditations: teaching us to wake up to ourselves, for the sake of happiness. Not all philosophy overtly calls for ritual meditation. For instance, epistemology, the study of how we know things, and eschatology, the study of how things end, involve conceptual investigation. But some philosophies, throughout history, have been about how we should live. Much life advice comes as part of a particular religion or politics. To indicate a philosophy primarily concerned with advice for living, I use the term "graceful-life philosophy." The important ancient ones were Epicureanism, Stoicism, Cynicism, and Skepticism, and the term is also useful for referring to the work of the Renaissance thinker Montaigne, and of any modern thinker who offers secular, philosophical arguments for how individuals should best live their lives.

Perceiving that worry and regret do us harm is a nice first step, but it does not, on its own, stop anyone from worrying or regretting. Montaigne wrote, "My life has been filled with terrible misfortune; most of which never happened." Spinoza wrote, "Repentance is not a virtue, that is, it does not arise from reason. Rather, he who repents what he did is twice miserable."[2] What these philosophers say is right, but is not the only thing that is right—which is to say, it is, in part, wrong. But this is how graceful-life philosophies, and many religions, try to change sad people into happy ones—by the repetition of well-formulated insights. Reading, thinking about, and even writing about the refusal to feel guilty *is the therapy.* The idea produces moments of relief from one's chagrin and opens up more ideas than it shuts down. Note that this method is not personal to you. The supposition is that we are all similarly plagued by jealousy, shame, and misapprehensions of our worth and that we can all use the same insights to heal ourselves.

The second big *know yourself* is Freud's. It is not entirely different from what Socrates and Plato were talking about, but it is different. Humanity was in a novel place in Freud's time. In the nineteenth century, having lit up the jungle with electric lights, we noticed that the violent chiefs, toothy women, and wild animals had relocated into the darkness in our heads. Law and asphalt had left them no place else. The hallmark of modern life is that the world is no longer a tug-of-war between various gods and people and animals. Now it is a tug-of-war between the gods, people, and animals inside each human being's mind: wolves and snakes, castrating father figures, and cannibal mothers. The work of Carl Jung, Freud's disciple and later disputant, illuminated parallels between our distressed modern minds and a set of timeless, powerful archetypes.

If the human world is not run by gods, it is run by human beings. If whenever you meet friends you are late, and keep them waiting and annoyed, you may not enjoy the experience, but what keeps you doing it is not a problem with your clocks or your transportation. Someone made you feel some way about being on time; perhaps you were forced to be too responsible too young, or perhaps you were humiliated to be left waiting and vowed to avoid it. If you can figure out why you feel better about yourself if you are late for meetings, you will likely be freer to change the behavior. Would that it were always so simple, but the great masters promise only measured miracles. The important realization is that the forces that "purposefully" stick out a foot and make you stumble are not demons out there in the world, but rather demons of your own mind, and there is a rhyme and reason to them. Don't supplicate, investigate.

Why are human minds so firmly affected by childhood experiences? I believe it is a part of what we call instinct when we see it in the animal world. Consider the lot of the young tiger. He has only a few perilous enemies to learn about, and his mother's horrified reaction to each one—the cobra, for example—sears into the tiger's mind. Think of some time when you were a child and your parents shocked and humiliated you with their angry, intense warning. Perhaps the story that leaps to your mind was about something truly dangerous, but more likely it was based on a socially variable issue, or something peculiar your parent worried about. If your mother once got smacked on the head by falling debris, or your father's youthful attempts at success were met by crushing failure, you are likely to be raised with a logic of hard hats and defeatism that may be all out of proportion to these dangers. For whereas the animal cub has a hundred lessons to learn and is thus taught a hundred lessons, human culture proposes millions of pitfalls, and the hundred lessons that were seared into your personality may not help you at all. Human beings come of age with so many maladaptive worries that, given a whole world to run around in, they usually pick something very narrow and do it over and over again. How do we pick? Well, Jung put it this way: "Nothing has a stronger influence ... on ... children, than the unlived life of the parents."[3] It is not only what your parents chose to do that influences you; it is also what they are aware of having missed. Children understand hidden messages. A mother who chose a rootless, independent life may feel betrayed when her daughter marries a doctor and settles in for a life of relative leisure, but mom may have given this choice such power that it had an unavoidable attraction.

As I persist in trying to explain the *know thyself* of psychotherapy, my text may feel busy with metaphors, but I think they are worth their trouble. Consider that we all have an internal empty field at birth, and as we grow, we experience shocks in certain areas of the field, which we respond to by building up a great pile of stones in that spot, to protect ourselves from being hurt again. As time goes on, the inner field grows crowded with stone mounds. Moving around in such a field requires inventive choreography; and that dance is what a personality is. A person with a lot of mounds is going to look pretty crazy when she tries to walk a straight line. When life circumstances change, the situation turns worse, since none of your long-developed shortcuts and coping methods work now. You crash into walls. The crashing makes you go to therapy, but you go to therapy looking for new shortcuts that will allow you to navigate your city of rock piles under these different circumstances, and what the therapist wants to do is bring you to the pillars and help you unpile the stones. There is nothing in the mounds to be scared of anymore, so if you can just budge the rocks, you will come to have free reign of your mind, and of the world, again.

Like philosophy, the work is strenuous and time-consuming. But what else were you going to do with your time? Maybe it will turn out that you fear death not because it is objectively scary that the inexorable thumb of the universe is headed down to squelch your living soul against the earth, but rather because you have not yet challenged yourself to dare to live, and you know it, and you have translated this wish for life into a surpassingly distracting fear of death. Maybe you miss your dead father so much because the old grump made it clear there were things you were going to have to prove to him about yourself, and now you will never have a chance to do so. Maybe what you needed and could get from him you already got, and all this awful mourning is about a mistake, his mistake, in not seeing you for the perfect and fulfilled little person that you were. Maybe if you come to see this, and forgive him, you can stop your heartbroken longing.

There is no easy way to find out what your problems are, because people do not come to therapy (or philosophy) to change their fundamental beliefs. That's what fundamental beliefs are; you don't even know what they are. People come to therapy because some adaptation they worked out with the world isn't working anymore. Some "symptom," or neurotic habit, that they used to be allowed is now getting in the way. They want to fix this immediate problem and are glad to replace this symptom with some other. The therapist wants them to find out what the symptoms are for. They resolutely do not. So they talk, and the therapist waits for something that seems unusual, and then asks more about that, until they can see that there is something particular about the way they see things. This second set of eyes is not the only reason you cannot do this alone. The other reason has to do with the tiger I mentioned earlier. What we take for instinct is imprinted so keenly because of the special relationship between parent and child. Families are isolated together, their members are intimate about much that is usually hidden, and these members have archetypical names and roles, as well as real ones. Our family is the place that sets our "instincts" and assumptions, and we need a similar role-heavy microworld to reset them.

There are brilliant philosophers and psychologists who themselves lead sorrowful lives. Somehow they make miserable choices, depriving themselves of love, money, and other normal comforts. Explaining to them what they are missing out on is like trying to get a dog with no legs to jump through a hoop by vigorously waving a steak on the other side. The problem is not motivation. Even the most insightful and motivated people cannot do the trick for themselves. You generally need someone outside yourself to help you change your mind. Consider a light version: You had an argument at home, then told the story to a friend and were shocked to find he or she thought you had been in the wrong. Remember how your hot anger changed to cold shame and you raced home? But it works on a deep level, too: you can shift your profound assumptions about the world, practically on a dime, with outside help. But it has to be done just right. This level cannot be approached by your friends; they know you are weird about certain things, but they don't know why, and they have learned that you won't tolerate discussion about these things. Defenses clang shut when threatened. That is why you need someone with some training. To return to one of my above metaphors: The piles of rocks and ramparts are not solid and real; they are inventions of the mind. No one else sees them, just the way you skirt around them. Because they are imaginary constructions, they are very effortful to maintain. It is exhausting to be heavily defended. People who are heavily defended may get a lot done in one or two areas, but they don't have balanced lives, because they are spending too much energy holding up their defenses. When we feel safe, when we feel we are with someone who basically agrees with us about the symbolic universe, we let down our defenses, confident that our companion understands the symbols that are usually walled up, and will act appropriately. The psychologist hangs out in your field with you long enough that he or she is allowed to make small suggestions about the symbols and whether they deserve the effort they take.

Psychotherapy is not just about pain. Just as you should know if your gun shoots slightly to the left of its sighting device, you should know if you tend to trust authorities or iconoclasts, for instance. You should know if the idea of the world as in decline seems right to you, or if believing in progress is your default setting. We each of us seethe over things that other people discount entirely. Aurelius said, "Life is opinion," and Shakespeare said, "There is nothing either good or bad but thinking makes it so."[4] This can sound like sophism or moral relativism, but it is more about the idea that with some self-knowledge, we can be happier. We have to find out why we think the way we think and what our assumptions are.

Socrates and Plato told you to unwrap your opinions on the world, and Freud told you to unwrap yourself. The third *know yourself* is less strictly associated with a doctrine or school, and, unlike the other two, is about knowing, not unknowing. It is expounded by anyone who has cultivated his or her own mind into a dependable and interesting friend, especially if he or she becomes lost and alone and discovers that with such a trained mind, no one can ever really be lost or alone. Boethius sat in a cell at the end of late antiquity, awaiting his death, and there wrote *The Consolation of Philosophy*, where Queen Philosophy came to him and helped him pass the time. Sweetly scolding his misery, she asks, "Do you really hold dear that kind of happiness that is destined to pass away?" All luck changes. What you need to be happy is a good conversation with your own mind, and prison is not merely unharmful to such a conversation, it might even be beneficial. Marcus Aurelius noted that having a self worth knowing was not only good for prison, it was also good for the absence of a vacation home. In fact, it was better than a vacation home: "Men seek retreats for themselves, houses in the country, seashores, and mountains; and you too are wont to desire such things very much. But this is altogether a mark of the most common sort of men, for it is in thy power whenever you choose to retire into yourself."[5]

Consider this influential testimony from another age:

> The chief reason for opening to every soul the doors to the whole round of human duties and pleasures is the individual develop- ment thus attained, the resources thus provided under all cir- cumstances to mitigate the solitude that at times must come to everyone. I once asked Prince Krapotkin, a Russian nihilist, how he endured his long years in prison, deprived of books, pen, ink, and paper. "Ah," he said, "I thought out many questions in which I had a deep interest. In the pursuit of an idea I took no note of time. When tired of solving knotty problems I recited all the beau- tiful passages in prose or verse I had ever learned. I became acquainted with myself and my own resources. I had a world of my own, a vast empire, that no Russian jailer or Czar could invade." Such is the value of liberal thought and broad culture when shut off from all human companionship, bringing comfort and sun- shine within even the four walls of a prison cell.

The speaker was Elizabeth Cady Stanton, and the speech is her famous 1892 address to Congress on the question of women's education and political rights.[6] I love the real-world example of a troubled girl that Stanton comes up with, clearly ripped from the headlines:

> The great lesson that nature seems to teach us at all ages is self-dependence, self-protection, self-support. What a touching instance of a child's solitude; of that hunger of the heart for love and recognition, in the case of the little girl who helped to dress a Christmas tree for the children of the family in which she served. On finding there was no present for herself she slipped away in the darkness and spent the night in an open field sitting on a stone,

and when found in the morning was weeping as if her heart would break…. The mention of her case in the daily papers moved many generous hearts to send her presents, but in the hours of her keenest suffering she was thrown wholly on herself for consolation.

This moment on the rock is maudlin, Victorian, and terrible. It's such an odd setting for claiming a need for education. I suppose that's why this speech is the classic that it is: no audience or reader ever expects Stanton to say this. They expect her to demand for women equal tools so we can work and live. Instead, the old lady is tired. She says, "To appreciate the importance of fitting every human soul for independent action, think for a moment of the immeasurable solitude of self. We come into the world alone, unlike all who have gone before us; we leave it alone under circumstances peculiar to ourselves…. We ask for the complete development of every individual, first, for his own benefit and happiness. In fitting out an army we give each soldier his own knapsack, arms, powder, his blanket, cup, knife, fork and spoon. We provide alike for all their individual necessities, then each man bears his own burden." We have friends in history, and we have a friend in our own mind. Without knowledge or education "the solitude of the weak and the ignorant is indeed pitiable." As Stanton put it, "In the wild chase for the prizes of life they are ground to powder."

Montaigne points out that, in the Bible, we were kicked out of Eden for eating from the tree of knowledge. Montaigne quotes Cicero rhapsodizing about the bliss of scholarship and asserting that he had learned to see the measure of things, and to be a generous man, from books about "the infinity of things, the immense grandeur of nature, the heavens in this very world." Such books, Cicero continues, "furnish us with means to live well and happily, and guide us to pass our age without displeasure and without pain." Cicero is claiming that knowledge and wisdom have made him happy: *knowledge,* meaning learned information about how the world works; and *wisdom,* meaning insight, generosity, and discernment. Montaigne can't bear the arrogance. He never describes himself as this sort of hero and seems to really dislike it when others do it. He writes of Cicero that "a thousand little women in their villages have lived a more equable, sweeter, and more consistent life than his." Montaigne cannot get over how much we do not know and how much we think we know. I doubt plowmen and village women lived happier lives than Cicero, but what is important is Montaigne's point that for all that we praise wisdom, it is not well associated with happiness. "Man's knowledge cannot make him happy," he tells us, because we are not equipped with the senses and the intelligence to understand much, so that even if there were happiness in knowing what is going on, we don't. Wisdom also fails because, in our attempt to offer wise solace, we find that thoughts, no matter how philosophical and pleasing, can only moderately affect the emotions. "Of the same sort is that other advice that philosophy gives, to keep in our memory only past happiness ... as if the science of forgetfulness were in our power."[7] But Cicero, Boethius, and Stanton make a good case for book learning as a source of happiness, like a friend.

Let me tell you one more thing about learning and happiness. In 1930, Bertrand Russell wrote that happiness is of two sorts, "plain and fancy." He explains: "Perhaps the simplest way to describe the difference between the two sorts of happiness is to say that one sort is open to any human being, and the other only to those who can read and write." Russell's gardener seemed siblimely happy in his eternal war against "they rabbits"; nevertheless, for Russell, "The secret to happiness is this: let your interests be as wide as possible, and let your reactions to the things and personas that interest you be as far as possible friendly rather than hostile." In the world, a full 82 percent of the population today is literate. In the developed countries 99 percent can read, and thus have access to job ads, letters, magazines, and possibly even the vast ocean of words and ideas.[8] Welcome to paradise. Seriously. Fancy paradise.

With all three versions of knowing yourself—Socratic questions, Freudian couch, and Boethian prison—the most difficult thing is that any sense of arrival must be preceded by years of difficult and often frustrating effort. Why is this so depressing? Doesn't the Ivy League med student know she is on a great path to a rich life? Why is she weeping into her locker at four A.M.? The process of becoming is a strain. Indeed, if it isn't agony some of the time, you are probably not doing it right. Plato's *Republic* is a proposal for a more perfect world. In it, when Plato metaphorically drags his fellows from their cave, the sun temporarily blinds them. They get used to it.

2

Control Your Desires

Montaigne lamented that for all their wisdom, some philosophers had lives that were famously blemished. The same can be said of many who have given reams of advice in modern America, from homemaking guru Martha Stewart to religious guru Jim Bakker. It's not just that their desires got them in trouble; it is that a person ought to be able to make decisions that are contrary to their desires but otherwise obviously the right thing. If one cannot manage that, and on such a major scale, he or she seems to be missing some common information about how to be happy. Maybe that's why such people spend so much time trying to make things a little better. Maybe the rest of us don't need festive centerpieces or weeping prayers because we have more happiness—as a collateral benefit of negotiating our desires.

The Buddha told us to master our worldly desires so that we can see the truth of the world around us. Our minds literally control our senses. We must, then, control our minds. If we don't, we'll have an enraged wild elephant on our hands: very difficult to manage. Essentially impossible. We should fear our desires "more than poisonous snakes, savage beasts, dangerous robbers or fierce conflagrations." Charmingly, he adds, "No simile is strong enough to illustrate this danger. But think of a man carrying a jar of honey who, as he goes, heeds only the honey and is unaware of a deep pit in his path! Indulge the mind with its desires and you lose the benefit of being born a man; check it completely and there is nothing you will be unable to accomplish."[1] Once he became the "Enlightened One" (which is what "Buddha" means), the Buddha did not leave humanity to go off and enjoy his new bliss. Instead, he made it his business to help as many others as possible. Helping people in this world is not the goal of Buddhism; the goal of Buddhism is enlightenment. But you need to control your desires and be a model of virtue in order to get there. And once you are there, one branch of Buddhism (the biggest one) insists that enlightenment will demand that you offer your life to humanity, so that virtue is a goal. But it still is not *the* goal. *The* goal is the blissful annihilation that is nirvana.

In Aristotle's idea of happiness, there is nothing higher than virtue. His idea of virtue came with weighty responsibility. If we could believe this, and follow through on it, virtue would likely make us happy. We have all had moments wherein we were aware of ourselves doing the right thing, and we felt happy. Also, we have worked hard in gloom and yet found ourselves happy at the end of the task. Yet something keeps us from making virtuous happiness a way of life. We get tired. Or just drawn back to the television. Virtue as a route to happiness cannot be discounted, but it has its difficulties. It was the later Hellenistic age that invented a lighter kind of happiness theory. The Epicureans defined happiness as joy in common pleasures: eating, drinking, sex, and friendship. Such pleasure-oriented graceful-life philosophy has an opposite set of limitations. Epicurus did not go so far as to say that virtue was for suckers, but he did say that politics and trying to change the world was largely a fool's errand. True enough, but some progress can be made, and for the individual a moral mission in life can be a very joyous thing.

Learning to control your desires, to just shut off wanting what you want, is a premier idea in all the graceful-life philosophies or self-help. Epictetus said, "Freedom is secured not by the fulfilling of one's desires, but by the removal of desire." Seneca explained his Stoicism as, most essentially, that reason "tames the madness of our desires." Marcus Aurelius, also identified with Stoicism, said you should be able to sleep in a palace one night and on the floor of a hut the next and be equally happy both nights. In his explication of a very different graceful-life philosophy, Epicurus wrote, "Do not spoil what you have by desiring what you have not; but remember that what you now have was once among the things only hoped for."[2] But despite calling for people to control their desires, Epicurus was a defender of pleasure. He said to be careful not to indulge in pleasures so much as to become ill, or to become dependent and thus a slave to it. But that was no reason to throw the baby out with the bathwater. Pleasure is great! "I do not know how I can conceive the good, if I withdraw the pleasures of taste, withdraw the pleasures of love, withdraw the pleasures of hearing, and withdraw the pleasurable emotions caused by the sight of a beautiful form." Philosophy and culture both go through cycles of preferring to control desires and cycles of preferring to indulge them. Both need attention, but there is something primary about learning to control your needs, and to put up with the grave whims of our brusque world.

Ancient Stoics said that happiness comes from living life in harmony with the universe—that is, that you should accept what life gives you and realize there is no reason to take any of it personally. They encouraged one another to show that they were okay with the universe by not minding pain or disappointment. Those who tout this plan too earnestly have always been teased for it. Cicero, for example, wrote that the Stoics were faking their calm in order to seem enlightened, and that when trouble comes, holding your breath and screaming inside is not all that much more noble than screaming out loud. Many wise people have rejected the idea of strictly controlling one's desires, but all recognize that some self-control is necessary to happiness. Part of the reason for this is that a good day is not always compatible with a happy life: TV and a beer is fun now, but good grades are a bigger joy, and they require some resistance to TV and beer. The big desires have always been food, wine, sex, revenge, riches, products, and fame. The danger—beyond fat, stupidity, syphilis, narcissism, taxes, clutter, and gout—is meaninglessness. These desires and the hunt to fulfill them feel meaningless because they are only intrasubjectively sensible: while you are in a fit of wanting, planning, and satisfying a desire—for revenge, say—it all makes sense. However, the moment after the gun goes off, or the moment someone snaps you out of your thrall, you can see that the whole thing is a small, dark, crazy mess, like a tangle of seaweed on the beautiful beach of a majestic continent. You somehow got your head stuck in the tangle of seaweed and haven't looked up in four years. When you do look up, out of either satisfaction or some shock, you will see that you have been wasting your time on something without any real merit or, worse, something that harms yourself and others.

The history of advice on food, wine, sex, riches, and products is discussed throughout this book, but we will here note that it is possible to shut down these desires almost completely, and that doing so is surpassingly liberating. Some people fall in love with controlling their desires, and as loves go, it is undeniably stable. And inexpensive. If you have never done it, you are missing out. But in the long term, for most people, shutting down your desires is not worth it. Moderation is tricky but seems to be worth the balancing act it requires. Even moderation takes a lot of learning to control what you want. People can teach themselves not to be ticklish, or to think their way out of an itch. Likewise, you can think your way out of wanting fancy shoes that cost two week's pay. Let us stay with products for a moment. Things can bring pleasure, but no one thing brings that much more than any other thing. People enjoy new things at every decent economic level, as well they might; there are enjoyable things to be purchased at every economic level, and one can save up and treat oneself at every economic level. There is no reason to want things that are entirely out of reach, though. Throughout history, the old and the wise tell the young and the foolish not to bother with wanting big diamond rings. They say: *Want what you have.* That may reject too much of what is amusing about being an American at the rise of the twenty-first century. Perhaps for us it is more reasonable to say: *Want what you can get.*

Of the prime desires that wisdom tells us to control—food, wine, sex, riches, products, revenge, and fame—the last two need attention here. Many people suffer from a desire for fame and may find some specific counsel from the philosophers useful. Almost all of us suffer from a desire for recompense for having been treated badly, and we could all use help rethinking the feeling. Consider fame first.

A great filmmaker gets a rotten review and mourns over it because, though it couldn't compromise the career's greatness, it is today's news. You, meanwhile, are not a great filmmaker, and though that can be a drag on some nights as you wait for sleep, you do not have to manage all those short narcissistic rushes occasioned by fame, and you do not have to cope with a terrible review all about what you've spent the last three years pouring your life into, right there on the front of the arts section of the *Times*.

Fame can be nice, but it is not the simple confirmation that it appears to be. Aurelius had numerous reminders for himself on this topic:

> But perhaps the desire of the thing called fame will torment thee. See how soon everything is forgotten, and look at the chaos of infinite time on each side of the present, and the emptiness of applause, and the changeableness and want of judgment in those who pretend to give praise, and the narrowness of the space within which it is circumscribed, and be quiet at last. For the whole earth is a point, and how small a nook in it is this thy dwelling, and how few are there in it, and what kind of people are they who will praise thee.[3]

If you are content with being a lawyer, nurse, teacher, or bridge builder for a living, there is no need for the hot pursuit of your own renown. So forget fame, say all the famous men. There's the rub. All our remembered bards of rejecting fame are, by definition, remembered. In some cases, for over two thousand years. What that suggests to me is that, like so many other things, you have got to live through fame to know why it is not heaven. They know that even the greatest star spends time bored, tired, or unable to sleep, and the periods of adulation may make mundane life more difficult to tolerate. The happiest you can be is how you are once fame has failed you and you have taken that in. The wise say if you want fame, try to get it, because it is worth something, but it won't be what you expected. The people you most wanted to impress may so resent your new status that you lose them entirely. You wanted to impress them because of how they were always impressing you, and that may be the only relationship with you they want. As for impressing the people of the future, Aurelius said, "See that you secure this present time to yourself: for those who rather pursue posthumous fame do not consider that the men of tomorrow will be exactly like these whom they cannot bear now."[4]

A big part of happiness comes from keeping friends and family and not hating those you work with, which means forgiving them despite your immediate desire, sometimes, to stay angry and hold out for recompense. There is no formula for knowing when relationships should end and sometimes it is right to keep a person in your heart but not see them anymore. But one of the most common ways in which people get estranged is that their feelings get hurt by something someone said in anger or behind their back. For this, you are better off letting go of your pain and anger and leaving the whole thing in the past. Not easy, but Marcus Aurelius offers nine rules for coping. They are:

1. Think about what your relationship to this person is, and whether you are his or her superior.
2. Think about who this person is "at the table, in bed, and so forth: and particularly, under what compulsions of opinion they operate."
3. Remember that they can't be happy about acting the way they do, and they must be acting involuntarily, through ignorance.
4. "Consider that you also do many things wrong, and ... even if you do abstain from certain faults still you have the disposition to commit them" and that you hold back only because you're scared to try or you're worried about what people would say.
5. Recall that "you do not even understand whether men are doing wrong or not," since we rarely have enough information to judge others.
6. Think about how "the man's life is only a moment, and after a short time we are all laid out dead."
7. Realize that no one else's behavior can bring shame upon you.
8. "Consider how much more pain is brought on us by the anger and vexation caused by such acts than by the acts themselves at which we are angry and vexed."
9. Try to get the person to see that people, like bees, ought to work together, and be kind to the person. "But you must do this not in irony or by way of rebuke, but with kindly affection and without any bitterness at heart, not as from a master's chair, nor yet to impress the bystanders."[5]

Then Aurelius offers "a tenth present": that "to expect bad men not to do wrong is madness," and that to see that they cheat, betray, and gossip about others, and imagine that they won't behave this way toward you, is irrational. You knew they did it to other people, and you tolerated it. Now that you know they talk about you, which you should have guessed all along, you should tolerate that, too. Bertrand Russell made a fine point of this, also, noting the indignant amazement of those who have learned of talk about them. "It has apparently never occurred to them that, just as they gossip about everyone else, so everyone else gossips about them."[6]

> When you hear that so-and-so has said something horrid about you, you remember the ninety-nine times when you have refrained from uttering the most just and well-deserved criticism of him, and forget the hundredth time when in an unguarded moment you have declared what you believe to be the truth about him. Is this the reward, you feel, for all your long forbearance? Yet from his point of view your conduct appears exactly what his appears to you; he never knows of the times when you have not spoken, he knows only of the hundredth time when you did speak.[7]

Letting go of the desire for revenge or recompense, difficult as it may be, is actually easier than getting satisfaction through exacting revenge. Spinoza explained that "[h]e who wishes to avenge injuries by hating in return does indeed live miserably. But he who, on the contrary, strives to drive out hatred by love, fights joyfully and confidently, with equal ease resisting one man or a number of men, and needing scarcely any assistance from fortune. Those whom he conquers yield gladly, not from defect of strength, but from an increase of it."[8] These are striking sentences. Even if you cannot find any other reason to meet rudeness with generosity, Spinoza has told us one that is hard to reject: you will win in the long run without almost any need for luck. The woman in your office who gave you the silent treatment for a year for no reason beyond envy, a fact acknowledged even by her friends, may have brought real stress into your life. You want her to suffer. But you don't really know what to wish for along those lines, other than massive success on your part or some fine shame for her. That would need luck. Instead, imagine that you choose to act with a strong good heart, reminding yourself that you can be above such injury and a model of virtue, and at the same time remembering that you, too, have faults, and that somewhere along the line you probably hurt this little monster's feelings. Maybe you didn't do much: she took your social discomfort for personal rejection. Or maybe you look like her judging mother. If you do win her over, it may be because she has learned something, even by watching you, even without knowing it. This way, when you win, she is more humane, not less, and you are more humane, and the world is more humane. Anyway, as Koheleth tells us in Ecclesiastes, "Be not hasty in thy spirit to be angry: for anger resteth in the bosom of fools" (Eccles. 7:9). With friends, you do not have to forgive them all. You can break away from someone who has hurt you, but if your head tells you to forget it but your feelings won't, try these forgiveness techniques, and think of it as a personal goal to work past your anger. As for family, you have to keep trying to forgive them, for your own sake. There is no way out of this one, because holding a grudge against family is exhausting and never ending, whereas by working to forgive them you can at least relax a little and expect progress. I'll add one more thing about holding a grudge over hearsay: never forget to consider the messenger. Why did this person tell you? With people who are neither friends nor family, you just want to learn Aurelius's wisdom against anger such that you do not care. As Epictetus said, "The essence of philosophy is that a man should so live that his happiness shall depend as little as possible on external things." It is good advice, but often the only thing we can do to flex the muscle is read and reread such advice and practice it through self-denial, hardship, and patience.

From the ancient world on, some philosophers have charged that trying to limit desires helps only so much, and anyway, it is no way to have a good time. Koheleth wrote: "Go thy way, eat thy bread with joy, and drink thy wine with a merry heart.... Live joyfully with the wife whom thou lovest all the days of the life of thy vanity ... for that is thy portion in this life, and in thy labour which thou takest under the sun. Whatsoever thy hand findeth to do, do it with thy might; for there is no work, nor device, nor knowledge, nor wisdom, in the grave, whither thou goest" (Eccles. 9:7–10).

Montaigne wrote that when he finds a voluptuous pleasure that "tickles" him, he does not let his senses steal the experience for themselves alone. Instead, he says, "I bring my soul into it ... to enjoy herself; not to lose herself but to find herself. And I set her, for her part to admire herself in this prosperous estate, to weigh and appreciate and amplify the happiness of it."[9] Not only does Montaigne take his pleasures, he stands up for them, which is certainly much harder.

> As for me, then, I love life, and cultivate it.... I do not go about wishing that it should lack the need to eat and drink, and it would seem to me no less excusable a failing to wish that need to be

doubled; ... nor that we should beget children insensibly with our fingers or our heels, but, rather, with due respect, that we could *also* beget them voluptuously with our fingers and heels; nor that the body should be without desire and without titillation.[10]

This is one of the funniest claims that life is for living. It is so convincing because he not only refuses to limit his joy in pleasure, he also brings his finer mind to the experience. Even more, he invents some pleasures just to show that if these existed, he would do them, too. Still, we need to know that we are in charge of our desires, and this takes practice and attention.

In our opulent (if often obscene) modern world, it seems severe to say no to it all. Yet one of the most important things we have to learn is how to cope with abundance and with our hunger for yet more abundance. Alcoholism is a complicated predicament, but it may be said that if you can have one drink, and stop, you can have another tomorrow, but that if you can't stop after one, you might have to stop entirely, for good. You have to be able to say no in any given *now*. We have so much food, and yet we want to be thinner than we have ever dreamed in history. A decade ago the headlines were about anorexia, now they are about obesity; both are all about control. The anorexic often is a girl with no power over anything in her life except what she eats, who cleverly manages to control the whole family by this tiny gesture of not eating. Obesity, too, is about control, of course. Sometimes it is the response of someone with more power than he wants; he makes himself submissive to his hungers. Sometimes it is a pantomime of rebellion; the dieter sighs *Screw it* as she tucks into her ice cream.

Being able to say *Not now* prevents having to learn to say *None for me, ever*. Today we have amazingly developed doctrines and techniques for learning to say no. The big lessons are these: A decision to change one's life has to take place in the present, because soon never comes. The present seems so small and meaningless that it seems too inconsequential to fret over. Confronted with a library of books, in this moment, you couldn't make a dent. Even if you read all afternoon, you're lucky to get fifty pages behind you. If you want to write a book, similarly, you cannot get much done in an afternoon: a few lines, a few pages. Yet the future will be constructed only by moments like this one, as they pile up in the past, just as the moments now piling up in your past are what you put into them, and nothing more. Ambition, education, dieting, sobriety, sexual abstinence, and physical exercise all require self-control and can all be overdone. Control your desires in the present up until the moment when you can say you have no problem with indulgence in them, on the basis of your recent past behavior. You have studied awhile? Get out and breathe some fresh air. Then go right back to controlling your desires.

The acclaimed short-story author Jean Stafford once said that "happy people don't need to have fun," by which she seems to have meant that a lot of what we do for fun is there to soothe misery. Stafford died at sixty-four of alcoholism and emphysema. Was it a good life? She won a Pulitzer Prize, an honor for which many men and women have sacrificed a great deal, most often without success; yet it is hard to find a happy period in her life until she married her third husband, and he died four years later. Modern commentators have argued over whether we should praise her and her literary generation for having "risked pain in pursuit of a more ambiguous pleasure: the pleasure of living fully" or whether we should pity her as lost "in the false romanticism of the tortured artist."[11] It may depend upon how much you value longevity and how much you enjoy parties, scotch, and cigarettes. Maybe "happy people don't need to have fun" accounts for only a few kinds of people. There are also those who are not happy and also don't get any joy out of Stafford's kind of fun. The singer-songwriter Sheryl Crow's 1996 pop hit "If It Makes You Happy" asked this intriguing question: "If it makes you happy / It can't be that bad / If it makes you happy / Then why the hell are you so sad?" The alternative to an unhappy man doing some questionable thing that makes him happy briefly is not only a happy man *not doing* the questionable thing. It is also the unhappy man who has not even found a way to make himself happy briefly. Don't take pleasures for granted. I love the Buddha, but it seems almost what I might call a sin to beat down our desires and their fulfillments. I should add, though, that the Buddha said you could return to pleasures once you reached enlightenment—that the technique he taught was a raft to go over a river, and that once you're on the other side, you do not need to drag the raft around with you. Still, he's willing to give up a lot to attain enlightenment. Whether you are or not, it seems like the thing to do is to learn to control your desires and, from that position, make decisions about which desires you would like to indulge.

3

Take What's Yours

The phrase *carpe diem* appears in the *Odes* of the Roman poet Horace (65–8 B.C.E.). The whole line is *Carpe diem quam minimum credula postero* ("Pluck the day, never trust the next"), so in a sense it is a call to remember death, but also it is the ultimate assertion to take what is yours. But what is yours to pluck? And how? Sometimes *carpe diem* means you should ditch work and enjoy a beautiful day, and other times it means you should seize the day to make a contribution to some grand project, however small the day's contribution would be. It all depends on what you will wish you had done with today when you get to tomorrow. The best way to know that is to ask yourself what you wish you had done yesterday. Seize the day, yes, but do not live as if every day is your last. Live as you wish you had lived yesterday.

Taking what is yours means more than *carpe diem*. There is also the matter of seizing your role. As we speak in languages, we act in roles. If you don't take on your role, things go wrong for everyone else. It is a bad doctor who explains to patients that a doctor is just someone who once went to medical school, and that a feeling of playing dress-up never quite goes away. A professor who takes no notice of her students' infractions does not serve them well. A driver who lets other drivers go when it is her turn to go disrupts the rhythm of the traffic. When roles change, from schoolgirl to bride, from new husband to father, from successful *macher* to the pitiable victim of some sorrow, it can be hard to adjust.

People don't know if they are good enough to take up certain roles, or can manage them, and they sometimes miss how much of this work the role will do for them if they will only take it on.

How about taking what is yours in the sense of things, power, and accolades? If you are born wealthy, the Buddha advises that you walk away from the money, just as Jesus advised that you give the money away. Both think you will be much better off without it. Most philosophers don't agree. Most philosophers instead propose that your role is to learn how to take up the ideal version of the role—in this case, the role of a wealthy life: to help others, but also to enjoy your wealth with elegance and grace. Ecclesiastes says there is no reason to envy the rich, because the troubles equal the benefits, but that if you are rich, delight in its pleasures. Marcus Aurelius wrote that this is the task: "With all your soul to do justice and to say the truth. What remains except to enjoy life by joining one good thing to another so as not to leave even the smallest intervals between?"[1] Isn't that great? The first part gives us our complete list of responsibilities in life. As long as you are doing that, be as happy as you can, even if that is very, very happy. Constant happiness cannot be achieved; he knew it as we all know it, but he did not argue against trying. According to Aurelius, if you find yourself in a position to join one good thing to another, you do not have to purposefully allow intervals of sadness, nor feel guilt, nor imagine some balancing portent of doom.

Aurelius says that one reason it doesn't matter how long you live is that this is not theater, that the whole is not the thing. Each moment is the thing. "The soul obtains its own end, wherever the limit of life may be fixed. Not as ... in a play ... where the whole action is incomplete if anything cuts it short; but in every part and wherever it may be stopped, it makes what has been set before it full and complete, so that it can say, 'I have what is my own.'"[2] There is important "forget death" stuff in this, but I am here highlighting the connected idea of allowing yourself to live. Let the length of life be of less concern, and look to being able to say, "in every part" of the life you have led, "I have what is my own." By the way, we think of Stafford as dying pitifully young, and Aurelius as living to a ripe old age (the white-haired Richard Harris played him in the 2000 film *Gladiator*), but in fact Aurelius died younger than she, on a military campaign. She was sixty-three, he fifty-nine. The low numbers remind us of the benefits of caution, but note also the ideals that they promoted in their lives. The last chapter of the *Meditations* begins thus: "All those things at which you wish to arrive by a circuitous road, you can have now, if you do not refuse them to yourself."

What if we take what is not ours? Consider that model of ancient heroic grandeur Alcibiades. The general was so handsome, charming, and cunning that he was forgiven for betrayal of his nation more than once, conned several kings out of empires, and talked whole armies into submission. His army once tried to sack a town but came too early, before his reinforcement armies were ready. When his unlucky group broke into the town's gates, the town's army had been forewarned, and encircled them with the intent to slaughter. Alcibiades shouted to them to lay down their weapons for they had been conquered by the Greek army. Surprised but impressed, they accepted his claim, or in any case put off killing him so long that reinforcements eventually arrived.[3] He did all this with sheer personal authority and, everyone says, a riveting handsomeness.

So is *that* what is yours to take? Maybe. But Alcibiades died young, at about age forty-five, and is certainly not remembered as a good man, when he is remembered at all. Wisdom and philosophy do not counsel their students to grab the world by the throat and pummel it into sycophantic and terrified submission. Or even to try to change things very much. The exception is in the struggle against oppression. Situations can damage people terribly without killing them, and growing up in danger makes it nearly impossible to become okay. Ever seen what happens to the self-esteem and ability to function of a youngster who has been sexually molested; or what happens to someone who gets beaten regularly? To fight such tragedy, philosophers—usually from the oppressed group—do encourage heroic behavior. Most other heroic behavior is seen as foolhardy by philosophers. In Dostoyevsky's *Crime and Punishment*, the "hero" is inspired by Napoleon to take what is his, which requires a first step, and the one he chooses is to get some money by a simple robbery. It works out badly. What he tried to take wasn't his.

Choosing what to take as yours is rarely obvious as a well-worn path, but neither is it entirely unmarked. Erasmus wrote, in *In Praise of Folly*, "For the most part happiness consists in being willing to be what you are." He has Folly add that "self-love has provided a shortcut." He's joking, but he still thinks it is best to love yourself, however you manage it.

You have to be willing to be who you are. When we say that inside every fat person is a thin person dying to get out, we assume that no overweight person likes the body he or she sees in the mirror, and that cannot be true. A chubby family in a chubby town without much media is likely to feel just fine. Why not own being exactly what you are? A man walking down the street with a friendly Great Dane has got to allow himself to be a man walking down the street with a friendly Great Dane. Otherwise, he will be miserable and ridiculous. If he gets into the role, he may meet women! All sorts of people. He will become known around town as the Great Dane guy, which should not get in the way of his being whoever else he is the rest of the time. You can look in the mirror and accept playing a young person and later, with equal vigor, look in the mirror and accept playing an older person. What nonsense to prefer one role to the other. The blow of most awful shocks is accepting the fact that your old life is gone and now you are this blind guy, or this girl in a wheelchair. People report that after they accept their new identity they feel a lot better. Speeches motivating people to succeed have been around for a long time, but they have never been the purview of the wise. Philosophers often praise scholarship; some praise sensual delight; some praise devotion to family, friends, and community. They never praise seeking worldly successes in money and power. So if the wise do not give motivational speeches for worldly success, who has given the wisest motivational speeches?

One answer is characters in plays. Someone too brilliant to propose certainty may still lend his or her brilliance to a brazenly certain literary character. There is no more rousing speech to fight gleefully against all odds than that delivered on the eve of the St. Crispin's Day battle in Shakespeare's *Henry V*. The king's armies are terribly outnumbered, and as they wait to launch the dawn attack, the men begin to lament their fates and fume against those back home in England in comfort. Henry scoffs, saying that for his part, he is glad their numbers are few: he is certain they are going to win and that the Englishmen now in their beds will be envious for all time.

> This day is called the feast of Crispian:
> He that outlives this day, and comes safe home,
> Will stand a tip-toe when the day is named,

And rouse him at the name of Crispian.
He that shall live this day, and see old age,
Will yearly on the vigil feast his neighbours,
And say "To-morrow is Saint Crispian:"
Then will he strip his sleeve and show his scars.
And say "These wounds I had on Crispin's day."
Old men forget: yet all shall be forgot,
But he'll remember with advantages
What feats he did that day: then shall our names
Familiar in his mouth as household words
Harry the king, Bedford and Exeter,
Warwick and Talbot, Salisbury and Gloucester,
Be in their flowing cups freshly remembered.
This story shall the good man teach his son;
And Crispin Crispian shall ne'er go by,
From this day to the ending of the world,
But we in it shall be remembered;
We few, we happy few, we band of brothers;
For he to-day that sheds his blood with me
Shall be my brother; be he ne'er so vile,
This day shall gentle his condition:
And gentlemen in England now a-bed
Shall think themselves accursed they were not here,
And hold their manhoods cheap whiles any speaks
That fought with us upon Saint Crispin's day.

Chills, right? It reminds any reader that in order to triumph over a seemingly insurmountable trouble, you first need to find yourself in a seemingly insurmountable trouble. It is a phenomenally important insight. In *Twelfth Night*, also in the mouth of a character, is a speech widely considered Shakespeare's *carpe diem:*

O mistress mine, where are you roaming?
O, stay and hear! your true-love's coming,

What is love? 'Tis not hereafter;
Present mirth hath present laughter;

What's to come is still unsure:
In delay there lies no plenty;
Then come kiss me, sweet and twenty!
Youth's a stuff will not endure.

Of course, these poems are just about battles and kisses, and even the wise will tell you that when it comes to battles and kisses, sometimes the only choice is a screaming, brutish plunge. We do not see eloquent exhortations to earn money and power until modern consumerism and democracy started insisting that self-interest is what drives community success.

From Bernard Mandeville, with his famous "Fable of the Bees; or, Private Vices, Publick Benefits" (1714), to Adam Smith's vision of capitalism as cooperation through competition, to what Max Weber famously named the Protestant work ethic, modernity has suggested that individual self-interest often serves communal needs. Get rich and your whole town benefits. Take what's yours and you are serving the community. The good grow rich. By the nineteenth century, ambition was lauded to a remarkable degree. Consider an excerpt from Rudyard Kipling's poem "If—":

If you can make one heap of all your winnings
And risk it on one turn of pitch-and-toss,
And lose, and start again at your beginnings
And never breathe a word about your loss;
If you can force your heart and nerve and sinew
To serve your turn long after they are gone,
And so hold on when there is nothing in you
Except the Will which says to them: "Hold on";

If you can talk with crowds and keep your virtue,
Or walk with kings—nor lose the common touch;
If neither foes nor loving friends can hurt you;
If all men count with you, but none too much;
If you can fill the unforgiving minute
With sixty seconds' worth of distance run—
Yours is the Earth and everything that's in it,
And—which is more—you'll be a Man, my son!

Well, in truth, what you'll be is an overly defended, tightly wound Victorian guy. Still, it is a fine motivational speech. Actually, come to think of it, "If—" is a motivational speech about war, too—and it encouraged the awful violence of the Boer War. What I want to point out is that all sorts of unnecessary pain is built into this motivational idea. First of all, what's all this about "winnings"? As we begin the last two stanzas of this famous poem, we are given the world as a high-stakes gambling game. Then, having lost everything, you are encouraged to "never breathe a word about your loss." Not good advice. Next, you are challenged to "force your heart and nerve and sinew" (force is a hard directive for such fleshly tissue). Look how "Will" (capitalized, alone along with "Man"!) is to be encouraged above all else, even to the point of wearing out its own heart. The next four lines all say that people are the same, that we should respect them all equally and not be vulnerable to any of them, ever. Also not great advice. What of the "unforgiving minute"? Wouldn't it be better to ask forgiveness from those family and friends you don't let in rather than proffering that power to a sixtieth of an hour? And must I run? (What if I'm in heels?) And what's this about "yours is the Earth"? Why do I want a whole planet? What about everyone else on it? Then, of course, there's that final triumph: masculine adulthood. On all counts, his ideals are open to debate.

If you think Kipling had some motivational excess, wait until you hear George Bernard Shaw. "This is the true joy in life, the being used for a purpose recognized by yourself as a mighty one; the being thoroughly worn out before you are thrown on the scrap heap; the being a force of Nature instead of a feverish selfish little clod of ailments and grievances complaining that the world will not devote itself to making you happy."[4] Again, such sentiments make most sense in love and war. Winston Churchill was rousing because his rhetoric had an appropriate context. We do not want to live our whole lives in blood, toil, tears, and sweat. It is a trance of value, and there are times when we need to be in a trance. Churchill's rousing advice is appropriate to sieges. It is part of our trance, too, though, and we are not being bombed, hiding in tube stations, shaking rubble out of our kids' hair and trying to find the courage to survive. We today are ridiculously goal oriented, rushing around like madmen and encouraging each other to reach for the stars, buy real estate, and add free weights to our workout. Most wisdom literature warns against taking the mundane world so seriously. Instead, perhaps, have a cup of tea, work a few hours on some long-term project, talk to your children or your neighbors. Seize the life. *Carpe vitam.*

Ralph Waldo Emerson's essay "Self-Reliance," of 1841, is a classic in the literature of taking what is yours, the gist being: "Insist on yourself; never imitate." It was a plea for Americans to stop slavishly imitating the past or the Old World and instead invent its own genius.

> There is a time in every man's education when he arrives at the conviction that envy is ignorance; that imitation is suicide; that he must take himself for better, for worse, as his portion; that though the wide universe is full of good, no kernel of nourishing corn can come to him but through his toil bestowed on that plot of ground which is given to him to till. The power which resides in him is new in nature, and none but he knows what that is which he can do, nor does he know until he has tried.[5]

You have to figure out what instrument your heart plays, and if that turns out to be something different than what you wanted, tough luck. Say you want to be William S. Burroughs but your words come out sweeter. Emerson says you must learn to respect this sound you produce, this sweet Burroughs, as a power new in nature, and you are the only source of it in the universe, and the only way we can find out what it does is for you to squeeze your heart and let the sweet Burroughs play. It is a brilliant piece of happiness advice. For those who have a reason to enjoy life as a grand campaign, Emerson will console you in your troubles. Emerson identified himself as a disciple of ancient Stoicism, so it is not surprising that one of his key points is to do what is in front of you to do, to play the part you find yourself inhabiting. But Stoicism rejected the starry-eyed fantasy of an individual attaining greatness. Emerson adds a motivational note, previously reserved for war and kisses, to the Stoic vision of the self: "Life only avails, not the having lived. Power ceases in the instant of repose; it resides in the moment of transition from a past to a new state, in the shooting of the gulf, in the darting to an aim."[6] There is no rest here, and this exhortation to action and constant self-realization was one starting place of modern self-help.

This call for constant growth means a constant rejection of your past.

We hardly speak of it today, but one reason change is hard is because it is a betrayal of your past. Of course, one can say, "Brooklyn works for me now, Manhattan was great for me before," but most of the time the effort of change leads us to describe the life we led before as a mistake. This means that when other people change (sober up, get divorced, leave town), it makes us nervous; it feels like an implicit critique, and in some ways it is meant to. Emerson says: "This one fact the world hates, that the soul becomes; for that forever degrades the past, turns all riches to poverty, all reputation to a shame, confounds the saint with the rogue, shoves Jesus and Judas equally aside."[7]

In tales from the Middle Ages, people were expected to make one serious conversion in their life, from bad to godly, or, at least as often, from good to fallen. Here in our modern world the motivation machine never stops pumping; people are never supposed to stop trying to grow and change, so there is a kind of instability of personal value. Emerson's contemporaries were uncomfortable with his idea of people in a constant state of self-realization. Today, we are all used to the idea that the people around us may announce at any time that they are trying to make a grand change, that they are finally going to become themselves, and that, yes, this is an implicit critique of who they had been, and perhaps of who you are now.

William James's *Varieties of Religious Experiences* is a book of philosophy, subtle and strange, as is all good philosophy, but meant to address the lived experience of real people, and meant to do so in prose that real people would find useful. As such, it mixes contemplation of timeless literature with contemplation of what was essentially the pop culture of 1902. As James reports, there was a tremendous fad raging called "the mind-cure." It was so popular that you did not have to be a devotee of the cure to know its details: "Its principles are beginning so to pervade the air that one catches their spirit at second-hand. One hears of the 'Gospel of Relaxation,' of the 'Don't Worry Movement,' of people who repeat to themselves, 'Youth, health, vigor!' when dressing in the morning, as their motto for the day."[8] James is not a drumbeater for the movement, and there is critique in his tone, but he also has a lot of respect for it. I think it is worth paying careful attention to where that respect comes from. He relegates one of his most important comments to a footnote. Here's the relevant excerpt:

One of the doctrinal sources of Mind-cure is the four Gospels; another is Emersonianism or New England transcendentalism; another is Berkeleyan idealism; another is spiritism ... ; another the optimistic popular science evolutionism of which I have recently spoken; and, finally, Hinduism has contributed a strain. But the most characteristic feature of the mind-cure movement is an inspiration much more direct. The leaders in this faith have had an intuitive belief in the all-saving power of healthy-minded attitudes as such, in the conquering efficacy of courage, hope, and trust, and a correlative contempt for doubt, fear, worry, and all nervously precautionary states of mind.* Their belief has in a general way been corroborated by the practical experience of their disciples; and this experience forms to-day a mass imposing in amount.

The accompanying footnote reads as follows:

*"Cautionary Verses for Children": this title of a much used work, published early in the nineteenth century, shows how far the muse of evangelical Protestantism in England, with her mind fixed on the idea of danger, had at last drifted away from the original gospel freedom. Mind-cure might be briefly called a reaction against all that religion of chronic anxiety which marked the earlier part of our century in the evangelical circles of England and America.

If you don't know what Berkeleyan idealism has to do with all this, you might do well to get hold of a 2004 film called *What the Bleep Do We Know!?* The film begins with physicists talking about quantum theory and then veers off from physics into a variety of self-help. The thesis is that because the world is less lawful than we had thought, we human beings make things real (such as they are) in the process of our perception of them, and therefore we ought to be able to influence our reality by thinking it into the shape we want it. In contrast to quantum mechanics, George Berkeley's eighteenth-century philosophy suggested that perception is all we can ever know. In response, Samuel Johnson announced, "I refute it thusly," and kicked a rock. The film will make many people want to kick a rock, because mixing physics and new-age mind cure ennobles the cure and insults the science. James reminds us that, long before quantum physics, Berkeley inspired similar fancies.

The origins of the mind-cure movement are given as decidedly transcendental, but their "Youth, health, vigor!" mantra gave them away as interested in worldly success. Still, James respects this mind-cure; he says it has benefited so many key people that it cannot be summarily dismissed. What makes it so important, he says, is that it is a necessary backlash against American anxiety, which he blames on American forms of religious devotion, particularly Evangelicalism. We often speak of modern New Age theory as spiritual, and as such, as a kind of religion, but it has always been an alternative to religion; indeed, it is proposed as a rejection of the rigidity and anxiety of much established religion. What made it so important to James still makes it important to us.

We are born into the game of modern life, and invited (politely or otherwise) to join in its activities: cultural, economic, familial, community, governance, science, and invention. The game is popularly understood as both awful and wonderful. Society is split between encouraging two extremes: (1) refusing this strange game of modern life; spending life on the sidelines, not having much of an impact on the world or on your own situation; and (2) accepting the game's terms as actual reality and invoking self-discipline, tireless efforts, and sacrifices in order to win big in one particular section of the game.

Refusal of the game is often touted as the smarter response, but it isn't. If you hear one part of your head vilifying the game as stupid or not winnable, you might consider whether this voice is trustworthy. The game of modern life—cultural, economic, familial, community, governance, science, and invention—is not entirely stupid, and you can win. The game is *somewhat* stupid, of course: business is corrupt and heartless; in culture, about half the time, the fix is in, and we viewers are offered the work of the connected instead of the talented; government operates more by greed and compromise than conviction; the paradise of winning is a false image: our sites of celebration, like Disney World, have a sinister other side. But on balance we live in an interesting time, and roller coasters are part of that. As lame as the game is, it is also a majestic continuation of human culture and we are lucky to be part of it. As for my insistence that you can win: you can. Almost everyone who really tries is able to see some of their goals come to fruition. Just showing up really is a great proportion of success. Ipso facto, showing up must not be easy, but if it is doable for you, expect to be rewarded. (The only additional secret is this: when you do show up, don't announce that you are better than everyone there, or worse than them.) Try different things, and press on further with what other people seem to like. It may be scandalous to say, but in figuring out which of your possible roads you should take, follow the praise. Do the activity that people seem to think you are good at, and take part at the level where people welcome you. (Even in matters of the heart this calculation is good: give your love to someone who shows signs of wanting it.) The point is to find a way to be engaged deeply with a bunch of parts of the game of life, but not to let winning (or any particular outcome) turn into an anxious, mind-closing obsession. Don't forget that you can take things by letting go of them.

The motivational part of *taking what is yours* is a modern addition. In the past, the injunction was not so much to better yourself and your situation as it was to accept yourself and your situation. We have reason to attend to both messages. Aurelius put it magnificently two thousand years ago: "Set thyself in motion ... and do not look about thee to see if any one will observe it; nor yet expect *Plato's Republic*: but be content if the smallest thing goes on well, and consider such an event to be no small matter. For who can change men's opinions?" Carpe diem, the favorite phrase of motivational coaches, is written in stone along the top left side of an old vaudeville theater that still stands in my Brooklyn neighborhood, built in 1904. *Carpe diem fugit hora.* "Seize this day: the hours flee!" On the right side it says *Ars longa, vita brevis.* It is to the brevity of life (and the endurance of art) that we now turn.

4

Remember Death

All the great graceful-life philosophies and all the great religions counsel people to remember death. Today we tend to want to forget death: our supermarkets make it seem like meat grows on white Styrofoam, almost bloodless, always pink. Dying happens elsewhere. We catch and release fish, then go have hamburgers. Human dying occurs in hospices, alone at night. Most people have not seen someone die, whereas, in past centuries, even young children were brought to deathbeds to witness a period of sometimes agonized dying, and then the much respected moment of transformation. This moment was as sacred and revered as the modern-day birth. Men of a century ago saw no births, or recordings of births, but they saw many deaths and recordings of deaths. Deathbed scenes were commonplace in literature and theater, magazine essay, and Sunday sermon. The idea was to educate people to have a good death and to help other people have one. For instance, an etching might show how the departing man or woman might reach up to God, and how the many guests should respond—for instance, that it was okay for some to look away, some to gaze with fixed attention, and for very young children to play beneath the chairs. Verbal descriptions of death offered a variety of models for how to behave in one's suffering, how to make peace and find forgiveness, and how to commend oneself to God. We have childbirth classes where you see and hear about a splendid array of births; they had essays and sermons that described an endless parade of deaths. In the future, classes for dying might look very much the same as childbirth classes, showing some breathing methods to manage pain and fear, and showing the partner, or "death coach," different ways of soothing the dying one. But instruction on death is not part of mainstream culture today. Indeed, studying it seems morbid, rebellious, adolescent. Valuing birth as a site for study and reverence and hiding death as a kind of profane, dark, secret, gross thing is, of course, a cultural trance. Cemeteries were often at the center of small towns and part of the life of the village; people picnicked there and daily visited their defunct friends and family. Even only a century ago, there were all sorts of customs around death that would seem bizarre and morbid today—for example, wearing a broach made out of elaborately twisted strands of your dead sister-in-law's hair. It is also true that until recently, much of history considered the execution of criminals to be an edifying and entertaining spectacle. Death denial may have reached its height at the middle of the twentieth century, a time when revelations about the Holocaust and memory of two world wars may have occasioned a general psychological shutdown on the issue. In a famous little essay of 1955 (it was two short pages!), "The Pornography of Death," sociologist Geoffrey Gorer pointed out that as the nineteenth century treated sex, so his time now treated death, and vice versa. Wrote Gorer,

> I cannot recollect a novel or play of the last twenty years or so which has a "death-bed scene" in it, describing in any detail the death "from natural causes" of a major character; this topic was a set piece for most of the eminent Victorian and Edwardian writers, evoking their finest prose and their most elaborate technical effects to produce the greatest amount of pathos or edification.
>
> One of the reasons, I imagine, for this plethora of death-bed scenes—apart from their intrinsic emotional and religious content—was that it was one of the relatively few experiences that an author could be fairly sure would have been shared by the vast majority of his readers. Questioning my old acquaintances, I cannot find one over the age of sixty who did not witness the agony of at least one near relative; I do not think I know a single person under the age of thirty who has had a similar experience.[1]

Gorer's guess about why this transformation occurred was that "belief in the future life as taught in the Christian doctrine is very uncommon today even in the minority who make church-going or prayer a consistent part of their lives," and that without such a belief, death and decomposition "have become too horrible to contemplate or discuss." An interesting theory, but what counted most was the observation itself: we were avoiding the subject. There were other calls for attention to death. Elizabeth Kübler-Ross's 1969 *On Death and Dying* generated a whole new vocabulary. Her five stages of grief gave us a way to talk about discreet parts of the experience of dying. She later explained that she did not intend her list—denial, anger, bargaining, depression, acceptance—to be understood as a neat series of events, just as you could write up the drama of falling in love as five stages but wouldn't expect a consistent one-to-one correlation when looking at real people's experiences. The fact that our culture has so mightily seized on her ideas, and for so long, tells us how desperate we are for some kind of script.

Gorer's comments about the deathbed scenes of theater bring to mind Margaret Edison's 1999 Pulitzer Prize–winning play, *Wit*, in which a brilliant scholar of the poet John Donne deals with her cancer and death open to the gaze of the audience. What makes that play work is decidedly not that the viewers had all seen this sort of thing in their real lives. The play is part of a change that took place in response to Gorer and Kübler-Ross, wherein the wise once again try to get us to stop averting our gaze. People seem to find this offer almost titillating. The HBO series *Six Feet Under* invited people to look at a corpse's journey from death to burial, and it was enthusiastically embraced by television audiences.

Even with the changes toward more direct conversation about death, the age-old advice to remember death, to keep it in the forefront of our minds for the sake of bettering the life we lead now, is still rather lost in our culture of youth, competition, and vigor. There is, however, one modern-day conviction about the life benefits of remembering death: the axiom that survivors of an almost-fatal experience are understood to be happier than other people.

The idea is that the cancer survivor lives every day in exquisite gratitude. What happens seems so fundamental as to suggest that there is a biological component to it, a change in brain pathways, or chemistry, or something of the sort. Let's call it "posttraumatic bliss." There are feelings in this life—good and bad—that cannot be conquered by intellect or force of will. Some are potentially blissful, like romantic love. Most are rough: the sudden loss of a family member, a violent personal assault, a brutal accident—or maybe something you did to someone else—soldier's remorse, for instance. Trauma flashes back. I am suggesting that almost dying can realign you in a way that is the positive incarnation of trauma: posttraumatic bliss. Throughout history we have tried to induce posttraumatic bliss by reminding ourselves of death. By this downright physiological interpretation, some aspect of self-induced posttraumatic bliss has been thought possible throughout the ages. People have tried to give themselves a jolt of the positive incarnation of trauma by traumatizing themselves with thoughts of death. Of course, the injunction to remember death is also intellectual; it is supposed to help you come to understand death in a way that won't distress you. But the way they tell you to study a skull reveals that it is more about shocking yourself into a post-near-death good mood. Consider a few brief and insightful attestations.

The Buddha wrote: "Of all mindfulness meditations, that on death is supreme." Koheleth in Ecclesiastes added this beautiful tangle of thoughts: "It is better to go to the house of mourning, than to go to the house of feasting: for that is the end of all men; and the living will lay it to his heart. Sorrow is better than laughter: for by the sadness of the countenance the heart is made better. The heart of the wise is in the house of mourning; but the heart of fools is in the house of mirth." The great Epicurus held that the "true understanding of the fact that death is nothing to us renders enjoyable the mortality of existence, not by adding infinite time but by taking away the yearning for immortality." The crowning insight comes from Aurelius: If, instead of fearing death, "you shall fear never to have begun to live according to nature—then you will be a man worthy of the universe that has produced you, and you will cease to be a stranger in your native land, and to wonder at things that happen daily as if they were something unexpected, and to be dependent on this or that."[2] Pay attention to living fully and you won't worry about death.

When Christianity arose, the wisdom to remember death was very well established. St. Gregory (329–388) was forever quoting Plato's advice to practice regular "meditation upon death." Ash Wednesday is a holiday devoted to remembering death. Each year, on the day after Mardi Gras, or Fat Tuesday, the priest takes ashes (from burning last year's Palm Sunday palms, and then mixing with a little olive oil to make them sticky) and rubs them on people's foreheads, in the shape of a cross. It used to be that clerics, who all wore what we think of as monk's tonsures (a shaved top of the head), would get their ashes in the tonsure. It must have been eerie to talk to a man, then see him turn to go and be confronted with this reminder that he was a man marked for death. Whether you get it coming or going, the minister of the rite marks you with ash and says, "Remember, man, that you are dust, and to dust you shall return." Catholics fast that day and don't wash off the smudge until after nightfall.

Plato would have loved it. He warned against poetic fantasies, so we generally think he would have found the promise of a Christian afterlife a little silly, but he loved the idea of enacting one's philosophy in ritual. The ancient philosophers always said that remembering death took active meditations and gestures. Ash Wednesday leads up to Easter, a holiday whose meaning is a fantasy of resurrection. But in and of itself, the holiday is philosophical and somber. Ash Wednesday arose from Christianity's two greatest sources: the Jewish influence came from the Jews' remembrance of death in a day of fasting on Yom Kippur. On this holiday Jews concentrate their minds on death, and rhythmically beat their fist against their heart. The Greek influence was in part as a supposed bad example: the holiday of Lent was set up as a specific refusal to join in the wild party of the Sacred Mysteries.

Christians also persisted in citing ancient calls to remember death. St. John Climacus, who wrote in the late 500s, advised people to "let the memory of death sleep and awake with you," and St. Benedict's guidebook for monasteries, known as his *Rule* (c. 530), advised monks to "see death before one daily." Eventually, Christianity formulated the idea of death as the time of reckoning, and of either torture or reward. Still, the fundamental message was that the contemplation of death was itself curative. Whatever you believe about your soul, your flesh is going to leave your bones. It is a potent notion. In medieval times, monks often had *memento mori* in their cells, objects kept because they remind one of death—usually, real human skulls or images of skeletons. The great

Renaissance sculptor Bernini designed a memorial for Pope Urban VIII in bas-relief: a tremendous bronze and marble skeleton, with awesome wings, holding a banner upon which the pope's name is written. The skeleton is in the midst of ripping the banner in two. Bernini found a spot on the dim church wall that was occasionally hit by a beam of light from a high window, and had the monument placed there so that it would be nearly invisible until it was revealed in brilliant glory. The image was so striking that people had it copied for their own monuments, and these too became known as *memento mori*. Much of the history of remembering death is comprehensible as a kind of posttraumatic bliss. It is not easy to invoke, but it works. It can make you feel mellow and happy.

In other cases, the call to remember death is more an intellectual argument than a technique for internal transformation, and its result is not mellow gratitude for the life you have but revved-up desire for more life. The thesis runs as follows: remember that you should fill your days with exotic action, because you could die at any time. In 1922 a French journal asked a range of noted people to answer a question—one of the main literary-magazine devices of the nineteenth century (another was the essay contest). This time the question was: "An American scientist announces that the world will end.... If this prediction were confirmed, what do you think would be its effects on people ...?" I love the gratuitous note that the scientist is American; it borrows our scientific clout to help sell the scenario, but it also feels a little like a premonition that American science itself may bring disaster. Anyway, here's how Marcel Proust responded:

> I think that life would suddenly seem wonderful to us if we were threatened to die, as you say. Just think of how many projects, travels, love affairs, studies, it—our life—hides from us, made invisible by our laziness which, certain of a future, delays them incessantly. But let all this threaten to become impossible forever, how beautiful it all becomes again. Ah! if only the cataclysm doesn't happen this time, we won't miss visiting the new galleries of the Louvre, throwing ourselves at the feet of Miss X, making a trip to India. The cataclysm doesn't happen, we don't do any of it, because we

find ourselves back in the heart of normal life, where negligence deadens desire.[3]

Proust then reminds himself and his reader that simple human mortality ought to be enough to get us in gear but, oddly, isn't. In his own life, Proust neither visited museums, nor threw himself at anyone's feet, nor even seemed vaguely interested in a trip to India. He liked to stay home in bed under heavy covers, eat stewed fruits, take tea and other drugs, and write his books. He did in fact die only a few months after he wrote this, not having gone much of anywhere. I do not agree with his own assumption that this was laziness. A lazy person could not even read the monumental À *la recherche du temps perdu*, let alone write it. The man made choices about how he wanted to fill his days. We know exactly what he would do if he had three months left to live: answer a magazine question, write his book, and generally take part in the great literary drama of humanity.

The idea that we need to remember death so that we will go to the Louvre almost sounds like there is a deathbed scorecard, or even an afterlife scorecard. There is not. There's a wonderfully silly country music song, "Not a Moment Too Soon," big in 2005, that claims that if we found out we were dying, we'd go skydiving, we'd go "Rocky Mountain climbing," we'd go "two-point-seven seconds on a bull named Fu Manchu." There is some truth to it. But remember, too, that we choose the things we do on a regular basis at least in part because we like doing them. Just because I'm dying I should suddenly feel like riding a mechanical bull? No. It is all metaphorical. We worry about death because we worry we aren't living. Living can be enhanced by doing exotic things, but that's not really so important. Noticing death may challenge you to do unordinary stuff. But noticing that you are especially wracked with fear of death—that should challenge you to think about what you wish were going on in your ordinary life.

Today *memento mori* are seen as a bit gruesome. If you are Catholic and you go to church or to your parents' house, you will see a statue of a dead or dying man. The rest of the time, like most everyone else, you live in a world that doesn't show you artistic images of a dead or dying person. The rituals of remembering death have largely disappeared, from communal chest beating to ash wearing. Yet despite the way natural death is concealed in the daily lives of Americans, images of death have snuck back into our lives and nearly taken over.

We live with constant images of death. The news photographs are not billed as *memento mori*, but with a little consideration of the idea, it becomes clear that they are. There have been inquiries into what we are doing when we photograph the affliction of others, and publish it, and purchase it, recycle the paper, and do it again. Susan Sontag's final book, *On the Suffering of Others*, is in part a recantation of her famous moral indictment of such photography of the early part of her career. She came to believe that looking at these pictures is more politically active than she had originally allowed. I am arguing that beyond politics, there is an old psychological need being filled. Given how often humanity has devised ways to look upon death in order to achieve happiness, and given how much we have sanitized our lives to free them of images of death, consider that the images of dead people that we see in the news, which is driven by our desires, are there to help us be happy. We shield children from news, but we still tell them to remember death. Think of classic fairy tales, and of Edward Gorey's doomed boys and girls. Think, too, of the dead parents of James of the Giant Peach, Lemony Snicket's Baudelaire children, and Harry Potter—all orphans. Children are going to think about death anyway, and it is much easier to think about death directly than it is to host the monsters that roam your head (and bedroom closet) when you try not to think about it.

Montaigne knew that thoughts of death could bring insight and liberation, but could be depressing in the short run. "It is certain that to most people preparation for death has given more torment than the dying." He blames philosophy for making us think of death so much and then rushing in like a hero to save us from our fatalism:

> Philosophy orders us to have death ever before our eyes, to foresee and consider it before the time comes, and afterward gives us the rules and precautions to provide against our being wounded by this foresight and this thought. That is what those doctors do who make us ill so that they may have something on which to employ their drugs and their art . . . They may boast about it all they

please: "The whole life of a philosopher is a meditation on death" (Cicero). But it seems to me that death is indeed the end, but not therefore the goal, of life.

Remembering death seemed to be something that intellectuals fretted over, while the less schooled managed the issue in a more natural, more relaxed way:

> I never saw one of my peasant neighbors cogitating over the countenance and assurance with which he would pass this last hour. Nature teaches him not to think about death except when he is dying. And then he has better grace about it than Aristotle, whom death oppresses doubly, by itself and by a long foreknowledge.

Only the educated think of death when they are healthy, "dine worse for it," and go looking for opportunities "to frown at the image of death." More-common people stand in need of no preparatory consolation and rather just take in the shock "when the blow comes."

Knowledge of death makes us human, and great knowledge of death can make us great humans. Still, I cannot help but agree with Montaigne that there is also the possibility of too much exposure to death, and too much acceptance of death. For some of us, it is hard enough to get our cleats into life. Remembering death is like switching to skates. The only historically defensible conclusion is this: Life does not seem like it is going to end. It is, though, and for your own happiness, you have to train yourself to accept it and keep it in mind. Only if you can cozy up to this peculiar fact can you be mature and happy, because otherwise you will be in worse trouble than normal if someone close to you dies; also because death lends life a gravitas and a sweetness; and finally, because the work of denying death will keep you as busy as a full-time job. But once you school yourself in the awareness and acceptance of death, you have to try to forget it again. Consciousness of death makes it too hard to invest effort in the present or the future. With knowledge of the sun's eventual expansion and absorption of the earth, the future is not what it once was. I am joking, as no one really needed science to get to nihilism. Over two thousand years ago, Koheleth said that everyone is forgotten. Want to know how forgotten we will be? I know you know all four of your grandparent's names, and that you may love them, and see them as your connection to your family's past. Good. Now list the names of your grandparents' mothers. I casually surveyed a bevy of Americans of greatly varied ages, and only a tiny minority could name even two or three of their four great-grandmothers. These are the mothers of people you have loved, spent days with, and possibly mourned. Koheleth was right. We are not going to be remembered.

Once we learn to remember death, how can we care about the future? The answer turns out to be romantic and moral; it is about our children most of all, but not only. Could I abide it if, after my death, someone burned the New York Public Library? I work my way back into existence from there. You want to be able to work your way back. Again, the way out of this happiness trap is to teach yourself to remember death, a long and laborious process, and then, though it will be almost as difficult, teach yourself to forget death again. As with controlling desire, remember that the Buddha said his method was a raft you use to get to the other side of the river; once you have gotten where you needed to go, you can stop doing the practice. Make yourself face death and become familiar with it. Effect within yourself a transformation. Seek out a state of posttraumatic bliss. But once you have done that, you have to firmly guide your attention back to life. Don't justify life and help it stand up to the paradoxes posed by eternity. Just walk your mind away from the dark edge of the beautiful springtime field and into its lovely center. Search a clover patch for a sprout with extra leaves, or roll over and look at the cloud-scattered sky.

Any time is all time. As he lay dying, the Buddha told his students not to grieve. He explained, "If I were to live in the world for a whole eon, my association with you would still come to an end, since a meeting with no parting is an impossibility."[4] Also know that Seneca tried to stop Nero from killing his rivals by telling him, "No matter how many you slay, you cannot kill your successor." Someone is coming, because, sooner or later, we are going.

Listen to our very motivated Shaw again: "Life is no brief candle to me. It is a sort of splendid torch which I have got a hold of for the moment, and I want to make it burn as brightly as possible before handing it on to future generations."[5] Of course, he's arguing with Shakespeare, through Hamlet. Hamlet sighed that life was so short we almost might as well extinguish it now, that it hardly mattered. "Out, out brief candle." The secret to the whole play is that his father's death, and the idea of killing to avenge it, have made death actual to him. He knows how an abyss yawns on either side of this tiny life, and these abysses make our stretch of life so tiny that the difference of years or decades loses all meaning. That's not how Shaw wants it. Not a "brief candle," he protests, but a splendid torch! We'll leave them comparing candles; it is a game even good men take to when they are scared of the dark. What's so bad about the threat of the dark, though? Neither Shaw nor Shakespeare must have ever been to high school or to a faculty meeting. They must never have tried to get home in the rain or had morning sickness. How could they have missed how slow life is? To those dogged by a nagging fear of death, it is like you are at a carnival with four lousy rides and a three-gag funhouse, playing rigged midway games, occasionally winning something and henceforth having to carry it around—and meanwhile, you spend a lot of time staring at the exit sign and worrying over eventually having to go home. Jeez, closing time is the least of your troubles.

As I mentioned earlier, I wrote this section, and really this whole book, as a way to develop an answer to Montaigne's challenge to wisdom. What I have come to believe might best be explained by a porcine romp through history. We follow the pig. In 334 B.C.E., Pyrrho of Elis joined the court of Alexander the Great. (Aristotle had finished his eight years as Alexander's teacher about a year earlier.) Pyrrho traveled with Alexander to India, where he studied with philosophers and ascetics. While at Alexander's court Pyrrho also came in contact with various Greek philosophers. When he came back, he became the founder of a new and powerful school of thought, Skepticism. In his hands, the gist of it was that every argument has a counterargument and that our senses and our reason frequently lie to us. Others would make more epistemological arguments based on Pyrrho's idea, but for him the point was that we should find Eastern calm in the realization that truth cannot be known. Once, when Pyrrho was at sea, a terrible storm tossed his ship around, and the passengers panicked and screamed. Pointing to a pig that was munching away on deck and looking at the waves without fear, Pyrrho told his fellow humans that the pig had the right idea. They all survived, and so did the story. Listen to Montaigne, almost two thousand years later:

But even if knowledge would actually do what they say, blunt and lessen the keenness of the misfortunes that pursue us, what does it do but what ignorance does much more purely and more evidently? The philosopher Pyrrho, incurring the peril of a great storm at sea, offered those who were with him nothing better to imitate than the assurance of a pig that was traveling with them, and that was looking at this tempest without fear. Philosophy, at the end of her precepts, sends us back to the examples of an athlete or a muleteer, in whom we ordinarily see much less feeling of death, pain, and other discomforts, and more firmness than knowledge ever supplied to any man who had not been born and prepared for it on his own by natural habit.[6]

Knowledge might be forgiven this failing if it were easy to acquire, but it demands supreme effort, over long periods of time, all the while promising consolation even as we can see, with our own eyes, that philosophers sometimes jump out of high windows, just like everybody else. They have meaningless weeping jags. We have heard of those who are rude to waiters, toady up to horrific political leaders, shove a chattering but innocent neighbor down the stairs, cultivate personal wretchedness, starve themselves to death, and allow their children to suffer deprivation and death. Some mock the poor and the worker, some have been racist, some have kept slaves, and most have been so vicious about women's minds and abilities as to constitute a gross inner failure in themselves.[7] Even the Buddha didn't give women full access to fairyland. With Montaigne, I have to say, "Wisdom, my friend, I am not impressed." Yet, the next thought must always be, "Ignorance, I am even less impressed; indeed, I am appalled." There is a big difference between those who try to be broadly humane and those who indulge and coddle their self-centered fantasies. Knowledge and wisdom are a lot better than ignorance and immaturity.

In 1861 John Stuart Mill wrote: "It is better to be a human being dissatisfied than a pig satisfied; better to be Socrates dissatisfied than a fool satisfied. And if the fool, or the pig, are of a different opinion, it is because they only know their own side of the question. The other party to the comparison knows both sides."[8] Remember this the next time someone asks why we support fine arts that do not quite support themselves. This whole pig-versus-philosopher debate is pretty hilarious, yes?

The upshot is that the wise can get wiser by watching the innocent. But that doesn't mean the wise want to switch places. What I have come to believe about all these paeans to wisdom flanked by all this despair about wisdom is this: everything has to be learned twice. In childhood we have ignorant happiness, and we must lose this happiness if we are ever to get beyond it. Repression is not the same as transcendence. Between these states of calm ignorance and calm knowing, there has to be some half-wise screaming. Some few people actually grow wise by acting wise, but most grow wise by acting foolish, by accruing a variety of experiences, by taking chances (and thus learning about chance, our constant companion), and by making errors. Voltaire's Candide said that we should cultivate our own gardens, but he concluded this only after he was no longer the naïf his name implies. By then, he had gone on many travels, seen a fine woman return from a voyage with but one buttock, seen the great philosopher of optimism lose one extremity after another until he was a mere nub of optimism, and seen kind men do as much terrible harm as cruel ones. What if all these characters had instead stayed home, cultivating their gardens, and had kept their buttocks, limbs, and innocence? They would have remained children. Or, as so many adult children are, pigs. Spinoza quoted Ecclesiastes' saying that whoever increases knowledge increases sorrow. But he was explicit in his opinion that, though both fools and wise people have happiness, they do not have the same happiness. As Spinoza put it, both a human and a horse have lust, but a horse lusts for a horse and a human for a human. Just so, Spinoza explained, the happiness of a drunkard is not the happiness of the wise. Knowledge and wisdom are worth it.

According to the great philosophers, your worst barrier against happiness is you, your own wrong thinking. Your four problems are these: You cannot see yourself or much about the world you live in. You are ruled by desire and emotion. You will not take your place or rise to your role. You are alternately oblivious to death and terrified of it. As such, your job is to master these four errors in yourself. If you do, you will be happy and more free to love, work, and play the way you wish you could. None of this comes easily; it has to be practiced a great deal, and it never works completely. However, there is no useful alternative to the effort. As Epicurus reminds us, "We must exercise ourselves in the things which bring happiness, since, if that be present, we have everything, and, if that be absent, all our actions are directed towards attaining it."[9]

The wisdom literature I have been discussing addressed itself to the question of how to live, but it is worth noticing, in closing, that the core of our lives is love and work, and these wisdom instructions also apply in these more specific realms. If you want your love and your work to be successful, actively apply yourself to knowing yourself, controlling your desires, taking what is yours, and remembering that love and work can both end abruptly—remembering to the extent that you cherish them both—and learn to let them flow. Actively applying yourself means doing something different in the service of these goals and seeking out information about what it is you need to learn. We acknowledge only some parts of the great store of happiness advice, and the most important idea that we have lost is that happiness takes study. Secular happiness requires the same kind of meditative work that religion requires: the solace works only if you rehearse, daily, weekly, and on special days, reading over and over the phrases and arguments that seem most persuasive.

Drugs

It is a modern myth that some mood drugs are good and some are bad. We are devoted to an anti-mood-drug rhetoric that does not match our behavior. We scold ourselves, as a society, for proscribing too many legal drugs and for indulging in too many illegal drugs, and many of us reject some drugs as bad. That still leaves a lot of drug-taking. Overall, our public rhetoric is mythically against drugs, and yet our individual lives include all sorts of intoxicants, stimulants, antidepressants, and other happiness drugs. It is powerful simply to realize that all these different drugs, the "good" and the "bad," are essentially the same: they are potions people use to get a little happy. Drugs can be dangerous; either the illegal or the legal ones may affect your health or turn out to be more than you can handle. But that is not enough to explain our attitude toward them. We need to see that we play down the pleasure aspect of the drugs that we allow, and that we do this for dumb reasons—in support of productivity and in defense of our war against the drugs we don't allow. At the very least it is worthwhile to think clearly about the drugs that we do take as happiness drugs under the cover of drugs to make us alert, asleep, or pain-free. It also seems reasonable to think about whether the "bad" drugs are really bad, and to remember that perhaps in a changed form these drugs might come back into the legal world again.

People have always used happiness drugs. Drugs that are now illegal were, at various times, used openly, as we use Prozac and caffeine. That seems odd to us, because modern technology, and modern markets, have made some of those traditional drugs a lot stronger. Also, consider that being involved in a normal life today requires more alert lucidity than ever before. We have appointments that begin promptly at 11:45 A.M. and not "near midday"; we deal in numbers and other symbols rather than in, say, people and cows. Another big explanation for why historical drug use surprises us is that we culturally disguise our own legal happiness drugs, calling them antidepressants, numbing agents, soporifics, or stimulants. Just what are modern "prescription" happiness drugs? We know what the pharmaceutical companies say on television commercials, and what a number of actors think about the question, and we also know that much of the medicine of mental health is either luck or bad luck. Still, savvy as we are, we are generally trapped in our era's assumptions and anxieties. Again I find only the terms "trance" and "trap" sufficient to describe the situation, and it takes a lot of effort to negotiate awakening and escape.

There are several reasons why it is worth the struggle to rethink this whole subject of happiness and drugs. One reason is to better understand what illegal-drug users are up to, given the perspective of several millennia. We also want to notice more about the way we use legal drugs today, perhaps to be more suspicious of them, but also to appreciate them more, as happiness elixirs. Once we notice that many drugs are, at least in part, happiness drugs, we also begin to notice the way a given society values happiness. As we have discussed, people want euphoria, but they also want their days to pass pleasantly, and for their lives to be good ones, overall. If a drug is understood as providing a nice burst of euphoria, how much risk to our pleasant days and good, long life will we tolerate from the drug? Or if the drug is understood to provide pleasant days, how much will we tolerate it if it steals euphoria from us? Or if it dampens our will to work and build? Happiness drugs offer a unique window into our individual and cultural balancing of these various desires.

People keep a lot of secrets about what drugs they take. Even if some people know what you take, and how much, other people in your life would probably be surprised. According to the National Center for Health Statistics, in 2002 and 2003 (the most recent tallying), for every hundred women from eighteen to forty-four years old, thirty-five had antidepressants prescribed, and seventeen had narcotic painkillers prescribed; from age forty-five to sixty-four, the numbers are fifty-eight and twenty-eight, respectively; and for every hundred women over sixty-five, fifty-three had antidepressants prescribed. For every one hundred men from eighteen to forty-four years old, eighteen had antidepressants prescribed, and eleven had painkillers prescribed; from age forty-five to sixty-four, the numbers are thirty and twenty-three, respectively; and for every hundred men over sixty-five, thirty had antidepressants prescribed. For every one hundred women and men over seventy-five, the numbers are sixty and thirty-two, respectively.[1] Sixty prescriptions for every hundred women over seventy-five! That's a lot of grannies on goofballs. Think of the people you know. Consider the proportion of people on one of these types of drugs. Is it consistent with the number of people whom you know to be taking one? If the numbers don't match up, your friends may be keeping this information to themselves. Some people brag about their gonzo Hunter Thompson moments, but, throughout history, many have been secretive about their drug use. Some people even hide their aspirin use. All this caginess makes it tough for the historian. It is like researching the history of cheating on your taxes, or the history of religious doubt. Still, some people have made public their private beliefs and habits and have called on other honest and intelligent people to "come into court, weigh up the evidence, and return their verdict," as Cicero said about the question of the existence of gods. It is worth the trouble to work to see beyond politics and propriety.

This section has four chapters. The first chapter addresses why some drugs are thought of as good and some as bad. The next chapter goes into the histories of cocaine and opium. I then discuss the way drugs have dropped out of religion but are yet understood as revelation. Finally, a look at drugs today brings us to the music show and drugs as solace. Some of this is imagined as what good people do, some as what bad people do, some as what good people do in secret. Again, my central point will be to demonstrate that these divisions are absurd. We believe them because we are lost in a trance of value, but that is all it is—a cultural trance.

5

What Makes a Good Drug Bad

Experts and parents used to say sports were bad for women. The idea was that exercise depleted the biological energies that women needed to be mothers. At that same time, experts and parents freely gave opium—often in the drinkable wine mixture laudanum—to girls and women at all stages of life, including pregnancy, nursing, and mothering. Today women are told that opium is bad for them and exercise is good, even in the final months of pregnancy. The change is not due to the progress of science. To the best of our knowledge today, exercise is associated with some risk in pregnancy and opium is not associated with much more risk. Drug users today might be more likely to also be poor, badly nourished, and overworked, but still opiate use is correlated only with a tendency toward low birth weight. The real difference is cultural. Victorians were most worried about women being too strong and rejecting the role of wife and mother. So girls and women were encouraged to be physically delicate and to alleviate their frustrations with drugs. Today it is commonly assumed that most powerful, world-wise women still want a husband and a baby, if later in life. So muscular girls are no longer our chief anxiety. It would seem, from the way we treat our pregnant women, that our chief worry today is whether this unnatural American life we share is bad for us.

Pregnancy is a hot-spot cultural moment: the stakes—the physical creation of the future nation—are high, so all the moment's anxieties

swell with import. We are rushed yet sedentary; we are artificially stimulated; we live in processed air and eat processed food and drink. So we
have our pregnant women go to prenatal yoga classes, get prenatal mas-
sages, and eat organic food. Why were they not doing this before, for
their own sake? Because pregnancy is a performance, a place for women
to act out whatever the culture thinks is good. For us, that turns out to
be about working less; resting; getting involved with Eastern relaxation
techniques (acupuncture and acupressure as well as yoga); getting big
doses of calcium, minerals, and vitamins; allowing a general plumpness;
and limiting the intake of anything that could be called a drug. Different historical moments protect their fertility in different ways. They
make different arrangements.

In these arrangements, the happiness of the pregnant woman is no
small concern—to herself or to others. She gets controlled, but also gets
the culture's acceptable treats or indulgences, whether that means drugs
that make you happy, or time to stand in a room performing Downward
Dog with a bunch of other women with enormous bellies. The pregnant
woman no longer seems to need her husband to run out and get pickles
and ice cream; that now seems like a performance to show that her husband was going to do what she asked when it came to the baby. This
performance is now enacted in attendance at birthing classes. Happiness drugs are not really gone, though. If our girl is on Zoloft when she
gets pregnant, her obstetrician's AMA-approved advice is to stay on the
drug. The terms used are: "These women get so much benefit from
these drugs that it outweighs the risk." Size of benefit outweighs risk.
Again, the benefit is not just to the woman. An upset pregnant woman is
no one's idea of fun. It is explicitly stated in the culture that a happy
mother is conducive to a healthy baby. Happy is good—still, notice how
sober, vigilant, and earnest this version of happiness is.

You may say that had they only been equipped with our science, the
nineteenth-century doctors who gave laudanum casually—to irritable
women, loud babies, and everyone else—would not have done it. I
assert, on the contrary, they would have used science to make a laudanum "safe" for pregnant women—that is, one that is not associated with
street drugs and that does not remove her entirely from the productive
world. Indeed, that is what they have done. Consider this: We know that
children die because of car crashes. We do not say, "Never let your child

in a car until the early teens when he or she fits properly in a seat belt." This would save the lives of many children, but we are not doing it. "Buckle up for safety" will do. I propose the following campaign: "Save a life! Never put your child in a car! Let's make it the law!" It is not likely to catch on. It would be too inconvenient. Everything about the way families live would have to change. With happiness drugs, an arrangement gets made based on various insistences of drug users, drug distributors, and the rest of society. Science can only assist in justifying whatever arrangements get made by everyone else. What risks are worth taking, what is seemly, and what penalties are worth incurring just aren't scientific questions.

The process of weighing benefits is cultural and historical. We are a historical moment, like any other, and our attitudes toward happiness drugs are such that we feel all sorts of guilt, pride, and pressure. These feel serious and real, but much of our attitude is a shared fantasy of societal needs and the playing out of cultural mythologies. Imagine that at some soon future time the birthrate of the United States (or that of a group within it) declines, and people begin to locate all their anxieties in a fear of national depletion. Wouldn't Prozac be seen as dangerous because it diminishes sexual feelings? Researchers might spend more time looking at any drug that tends to increase sexual activity, and as such, they might become conscious of other health benefits of the drug. Science isn't outside of culture, because even the questions we ask are culturally determined. We don't go looking for, say, an intelligence correlation with those who have large second toes. Nobody has a vested interest in the answer, so there is no interest. If people become interested, studies are done, and the ones that serve people's interests are found interesting. Modern prescription happiness drugs, like historical happiness drugs, are made of materials we found in the forest and then heated, strained, fermented, dried, or more profoundly doctored. Right now they seem like medicine, like authority, while illegal drugs (and alcohol) are very often understood as rebellion and freedom. There is a good girl who takes Xanax but looks down on potheads, and a bad girl who rejects Xanax because taking it would mean there is something wrong with her but happily does meth.

Tea and coffee are not today thought of as happiness drugs, but that is odd. Ornate rituals developed around tea on the opposite sides of the

world, in Japan and England. In these countries and in others, especially China and India, we find paeans to the leaf as a happiness drug—to a
degree that can make you wonder if everyone is getting the same tea. In
the eighth century, the tea sage Lu Yu wrote, "Tea is not only a remedy
against drowsiness. It is a way of aiding men to return to their sources, a
moment in the rhythm of the day when prince and peasant share the
same thoughts and same happiness while preparing to return to their
respective fates."[1] The happiness provided by tea should not be underes-
timated. In 1757 Samuel Johnson called himself "a hardened and
shameless tea-drinker, who has, for twenty years, diluted his meals with
only the infusion of this fascinating plant; whose kettle has scarcely time
to cool; who with tea amuses the evening, with tea solaces the midnight,
and, with tea, welcomes the morning."[2] A little over a century later, in
1865, Prime Minister Gladstone wrote of tea: "If you are cold, tea will
warm you; if you are too heated, it will cool you; if you are depressed, it
will cheer you; if you are excited, it will calm you." Tea was for amusement, solace, and cheer—within a schedule of hard work—and the
overworked nineteenth-century Englishman brought love of tea to an
extraordinary height. Ladies at temperance meetings on both sides of
the Atlantic sipped tea to refresh themselves for another diatribe against
the wastefulness of alcohol. They even fetishized their drug paraphernalia of fine tea sets and spoons, laying out their tea works in breakfronts,
like a teenager with a water-pipe display. When Britain's industrial and
financial dominance began to fade, her people kept drinking tea, but
the love affair has become quaint. Its image was exotic at first; later it
was fuel for industrial dynamos. Now its image is gentle and traditional.
The English still drink an amazing amount of tea: according to Britain's
Tea Council, the British are the world's largest per capita consumers of
tea.[3] Yet the paraphernalia of tea drinking is often antique, its niceties
the ways of an earlier era.

Coffee vendors have reconstructed the meaning of the drug in every
age. The French café was the center of the later Enlightenment and of
the democratic revolutions. People drank espresso there and got a lot
done. Early twentieth-century America treated the coffee shop, with its
weak American brew, as a place to rest. Remember the "coffee break"?
Coffee was recharging you, but it was also relaxing you. That had to do
with the nonwork setting of the coffee shop (no laptop computers, no

kids) and the low caffeine levels of the drink. You wanted a small cup of weak brew because you wanted the waitress to come back and pour you another hot cup, and chat for a moment. Someone who comes to you, unbidden, with a hot drink and also offers pie is a mother, or a sister, or a sweetheart. She takes you seriously; she watched you to see when you were finished. Terms of endearment were common forms of address among strangers. This was about home and rest. Present-day coffee bars provide strong coffee again and offer themselves as offices and child-care centers: bring your own computer or toddler. At most of them, like in a nightmare of bureaucracy, the person who brings you your drink is not the person who took your order. This coffee place is not to be mistaken for a home: it is a place of efficiency. The method often slows things down, but it looks like an assembly line of individual service, and that's what felt right about it as it expanded across America and showed up on every city block through the 1990s. The values of the modern coffee bar are clear: you can stay forever if you're working, but even with a purchased chai latte on the table in front of you, you are not supposed to take a nap.

The overworked twenty-first-century American loves coffee. Not only do we drink it; we think of new ways to sell it, buy it, and make it. Our love of coffee will be part of our legacy: in our time, Americans worked long hours, and used strong caffeine derived from the coffee bean. We always speak of caffeine's effects as waking us and helping us to work. We rarely speak of it as a happiness potion, a calmative, or a break from work. Yet a less overt part of the cultural conversation indicates that we do still use the drug this way, especially as a happiness drug. Think of how we act out our coffee experience: longing for it, sighing with pleasure when it arrives. High-caffeine drinks are bought in cans also, especially by the large consumer block of drowsy high-school and college students. The strongest canned caffeine drinks are imagined for the industrious, the hyper-alive risk taker, and the teenager wanting a buzz. We allow it because we need these kids to be awake in class. Red Label, a whiskey, is not sold in college food courts, and Red Bull, a caffeinated drink, is. They are both powerful drugs. This is not about health; it is about culture.

Why did the people of the 1890s like opium when we like Lexapro? We drive cars. On a horse, even if you are very drunk, you are unlikely to accidentally run at full gallop into a brick wall. Clearheadedness has never been so important in all of human history. Goofiness is to be avoided. We are maddeningly productive, and so, since there is often no test to determine which person is actually being productive, we do not want to seem stoned. Our successes are often somewhat dependent on the efforts of people who are strangers to us. We want them sober. We mostly want clearheadedness in our school-bus drivers and our commercial-jet pilots. For our own sake, to function in the world, we cross busy roads, handle money, show up exactly on time for meetings, have professional conversations, and try not to be asleep when the train arrives at our stop. Not drooling is a big priority for us. What makes opium a bad drug and Zoloft a good one has a lot to do with fogginess. We think of the good happiness drugs as fixing something medically wrong with the depressed person, and of the bad drugs as masking the problem with a gauzelike inebriation. It is more accurate, and useful, to note that the degree of inebriation, or gauziness, it causes is one of the main differences between a good and a bad drug, and not whether one heals and the other masks. We have no evidence of that; it is an interpretive idea. Rather, the gauziness is a problem for drivers of machinery and yeomen of industry—that is to say, all of us.

The other part of the good or bad reputation of a happiness drug is the risk of dependence, but that is not as straightforward as it might seem at first. Many of today's celebrated prescription happiness drugs are difficult to stop using. It always takes society a while to figure out which drug can be put down and walked away from. Starting in 1898, heroin was produced by Bayer, and so named because it was of heroic help for coughs. It may seem to us that the cure was a bit extreme for the problem, but in the late eighteenth and early nineteenth centuries, Europe and the United States were savaged by tuberculosis. A young man or woman might one day start coughing blood, look pale, sometimes feel weak, and then cough more blood and eventually die. It sometimes happened fast, but he or she could walk around like this for ten or twenty years, infecting other people, trying not to get too depressed. Heroin helped immeasurably. When it first appeared it was touted as having all the benefits of opium with none of the addiction. People were told to eat or suck on the heroin, and so they didn't get as high or as addicted as they would later by injecting it—but it was still heroin, and it provided a burst of well-being. There were concerns, but the impulses that ruled the day were to benefit the suffering, who were doomed anyway, and to get them to stop coughing for the sake of everyone else. My question is whether the infamy of some drugs is primarily due to their addictive properties. It is not entirely silly to note that we are dependent on food and we manage the problem, though we have to spend much of our time at it, and such happiness is nothing to dismiss just because it is difficult to arrange. Also, not all illegal drugs are addictive: generally speaking, there are no withdrawal symptoms when a pothead stops lighting up. Note, too, that it can be relatively easy to deal with addiction if the whole of society is in on it with you: if it wants to, society can allow you your time, supply the drug, and have a few all-night kiosks available in case you've planned badly. In the Middle Ages and today, places in the Muslim world make it possible to get opium if you need it, and even in Catholic France on a Sunday you can find a place to buy coffee and cigarettes.

Which drugs are seen as worth the risk and which disdained? One big answer is that drugs become associated with either the upper or lower classes and this association sticks for a while, and then shifts. For a while in the modern West, poor people snorted cocaine, which seemed boorish and grotesque to the rich, who injected it; a century later the rich would be the snorters and the poor would smoke the stuff, in the form of crack. A report from Bengal in the late nineteenth century explained how opium was for the rich and marijuana smoking was for the poor, though at celebrations both rich and poor, men and women, old and young—all partook in a marijuana (*bhang*) and sherbet concoction famous for inducing hilarity.[4] Why do we have such a strong feeling that using happiness drugs of any sort at all is a little "bad"? As drug historian Richard Davenport-Hines has put it, drug campaigns that arose in the early twentieth century are responsible for characterizing drug use as "naughtiness." It is a cultural trance at this point, an incredibly tenacious one, such that even when we are intellectually aware of the history of drugs, we still respond emotionally with our era's codebook of concerns. How to break the trance? We need immediate examples to strengthen our guess that these anxieties are half hobgoblin.

Humanity takes happiness drugs to quiet the pain of bad memories, to escape a bad situation, to increase sexual desire, to shake off boredom and lethargy, or just to feel normal, to stop crying so much. We have also counted on a number of powerful intoxicants as protectors from disease. This wasn't just a coincidence, or just an excuse to take drugs: the ability of these agents to affect the body was taken as evidence of their efficacy at other tasks. Civil War soldiers took opium to prevent malaria and diarrhea. It was opium, so they were also taking it to get high, but the experience is not an easy one to classify. It wasn't "naughty." Now we label some happiness drugs as "good person" drugs and some as "bad person" drugs, and you might take a whole lot of legal drugs before noticing that taking them is a bad idea for you. Or take a few illegal ones and feel guilt. Or take the legal kind and feel guilty anyway, because you associate all drugs with naughtiness. You might avoid modern, prescription happiness drugs because of the hubris of doctors; that is to say, you prefer illegal drugs because they seem like they are for pleasure and freedom, whereas the prescription ones seem associated with control. The preference for good-person drugs or bad-person drugs means that one either trusts experts and authorities, or doesn't. Both sides, remarkably, make this point by using their own bodies.

We often indict our own culture for supporting an unusual amount of mood-altering drug use, but a glance at the history of drugs shows this perception to be false, and not a little unfair to ourselves. Sobriety is rare. All cultures have had favorite ways for people to take in some chemical in order to feel better. In every life there is useless pain, fear, and shame; there is anxiety; there are days when it is hard to do what has to be done; and there is boredom. People use drugs to help, and they always have. The ancient West had fermented drinks, opium, and some drugs that appear to have been hallucinogenic fungi. The ancient East also had fermented drinks and soma, the psychedelic euphoric botanical that was at the center of ancient Hinduism. We find references to drug use in Europe throughout history, certainly pot smoking. In fact, we have discovered special pot-smoking braziers, where people would make a little fire of dried pot leaves and sit in a small room together and breathe the smoke. We have found these braziers, with cannabis in them, dating back to the origins of agriculture and continuously straight through the medieval period. In the ancient Indian text *Artharvaveda*, cannabis is one of the herbs that "release us from anxiety." Yet the question of the use of happiness drugs is more complicated to report on than that, because there were a lot of ways to get high. Medieval Europe was a starvation culture. Every generation knew famine. In bad times, when there was nothing else, people ate the inedible: people were found dead of starvation, with grass in their mouths. People ate bark when there was nothing else; they ate bread made mostly of sawdust. Writers described finding corpses of people who had gnawed away their own hand or knee. Before it came to that, people would eat anything, which included poppies, hemp-seed bread, any mushroom or berry, and all manner of spoiled food.[5] Historians have noted that there is no question that medieval men and women could find drugs. The question is: were they able to avoid them?

If there had been a great deal of drug use, we would expect to find some complaint about it, even if it fit into a very different social meaning; and, indeed, there is. For example, hemp—marijuana—was banned by the Church in medieval France. Rabelais, born sometime after 1450, wrote that his giant Pantagruel was named for an herb, which he called Pantagruelion, which was marijuana, right down to the leaf formations. He devoted three entire chapters to its praise. The giant smoked it constantly, and ate hashish-laden Turkish delight, traveling with a colossal supply of each. In the final book, Rabelais reveals that Pantagruelion is hemp. It is surprising to us, but perhaps it should not be. It seems we have been ignoring much evidence and attestation. William Salmon, writing in 1693, says that cannabis seeds, leaves, juice, essence, and decoctions could be bought at any druggist's shop in Europe, and that cannabis was a widely used medicine. Drugs for happiness are discussed in texts from the Renaissance and early-modern period (usually defined as extending from the end of the Renaissance to the start of the Enlightenment). Writing of the many pharmaceutical cures for melancholy in 1621, Robert Burton explained, "Every city, town, almost every private man hath his own mixtures, compositions, recipes."[6] Also writing in the early seventeenth century, the poet John Donne recorded casual opium use in the beloved "Holy Sonnet X." The poem insists that there is an afterlife and therefore we ought not be worried about death, but it also murmurs that we are tired and need sleep. As rest and sleep are cherished in life, why not look warmly at an end to life?

DEATH be not proud, though some have called thee
Mighty and dreadfull, for, thou art not so,
For, those, whom thou think'st, thou dost overthrow,
Die not, poore death, nor yet canst thou kill me.
From rest and sleepe, which but thy pictures bee,
Much pleasure, then from thee, much more must flow,
And soonest our best men with thee doe goe,
Rest of their bones, and soules deliverie.
Thou art slave to Fate, Chance, kings, and desperate men,
And dost with poyson, warre, and sicknesse dwell,
And poppie, or charmes can make us sleepe as well,
And better then thy stroake; why swell'st thou then;
One short sleepe past, wee wake eternally,
And death shall be no more; death, thou shalt die.

What he says here is: Death, you think you are so powerful, but the
people you think you kill don't actually die, and you cannot kill me,
either. Death, you think you are so great at putting people to sleep, but
poppies and other potions make us sleep, too, "and better than thy
stroake." Death is in part vanquished because of the goodness of poppies
in comparison. We think more along the lines of Shakespeare's sonnet
on love that says "Love's not Time's fool," and we sort of pretend that is
what Donne is saying—that love and life transcend death—but
Donne isn't. Donne is saying something about sleep, and poppy trips,
and the pleasure of being finished.

It often sounds as if historical personages were using drugs in ways
quite similar to the ways moderns use them. Still, the reality must be
something wholly other; not like us, but also not like the magically drug-
free thousand years that we usually imagine. States of mind were imagined differently; medieval culture had its own imagery of health and
disease, religious "conditions" both blessed and damned, varieties of
madness, and levels of village idiocy. What we would consider "drug
use" was mixed into these in ample supply. Especially with addictive
substances, it is easy to mistake withdrawal reactions for symptoms of a
disease, and the drug as a restorative.[7] Past historical eras left records of
diseases that we do not have today, and some of these were deemed
responsive to opium. Some of them may have been opium withdrawal.

Similarly, the Victorian age set up a gender division that was too extreme for many people, and illness was a refuge into which patients could flee—and to which doctors could resort for an explanation of problematic feelings and behavior. Women were often diagnosed as hysterical if they were sexual, energetic, garrulous, loud, or boldly emotional; in acute cases, there were paralyzed limbs or blindness. Nervous exhaustion, often called neurasthenia, was often diagnosed in men who were tired, emotionally sensitive, and subject to the occasional collapse. Illness is rarely flattering, but if the status quo was impossible for someone, the choice got made. It is never fun when you are considered sick and people can boss you around, but one perquisite for going this way was (and is) the drugs. Consider, too, that the diagnosis of nervous exhaustion was well positioned to be a cover for, or an interpretation of, a man whose maintenance dose of some drug had gotten out of hand. Some theorists such as the French philosopher Théodule Ribot, considered drug addiction to be essentially the same thing as nervous illness.[8] In an age of "Man for the sword and for the needle she," as Tennyson put it, hysteria was the label for a loud girl who had a metaphorical sword and wouldn't quietly sew. I am proposing that nervous disease was sometimes the diagnosis for a man with a needle. Of the hypodermic variety.

I will offer a last thought on the general question of what makes a good drug bad. The Mormon Church, the Church of Jesus Christ of Latter-day Saints, was formed in 1830 (Joseph Smith had his first vision a decade earlier), and part of the doctrine was to abstain from alcohol, coffee, and tea. These avoidances were not mandatory for members until the twentieth century; indeed, it wasn't even until 1906 that Mormon temples started using water instead of wine in their sacraments. When the Mormons made their big trek to Utah in 1847, they provisioned themselves with coffee, tea, and alcohol, and if you check their medical recipes of the time, they include both brandy and laudanum. But after the Mormons got to Utah they began rejecting caffeine more strictly. When the Native Americans introduced them to a piney-tasting tonic made from a desert plant, they liked it so much that it soon became known as Mormon tea. It was ephedra. According to the FDA Web site, "Its principal active ingredient is ephedrine, which when chemically synthesized is regulated as a drug. In recent years ephedra products have been extensively promoted to aid weight loss, enhance sports performance, and increase energy." Studies have found that the drug raises blood pressure and otherwise stresses the circulatory system. These reactions have been linked to significant adverse health outcomes, including heart ailments and strokes—some fatal. In response, the sale of ephedra has recently been prohibited or limited. Mormon tea is thus now understood to be a lot stronger and more dangerous than coffee or tea, but Mormons, having never had coffee or tea, did not know this. Careful whom you scold, friends—and take both your guilt and your pride on such issues with a sip of the Great Salt Lake.

6

Cocaine and Opium

The story of cocaine can tell us a great deal about happiness drugs, and why they are sometimes considered to be a natural part of life, healthy and beneficial, and sometimes rejected by all upstanding citizens. Cocaine users speak of the drug as a happiness high that allows you to get a lot of work done. Unlike marijuana, which is often cited as relaxing the user's work ethic, cocaine is a drug of productivity. Unlike amphetamines, cocaine is associated with euphoria. Cocaine, then, is a very interesting subject through which to ask questions about what our society tolerates. Everyone wants to be happy, but what is happiness? Can true happiness be drugged happiness? You were happy today. Does the fact that you had two cups of strong coffee and a dose of over-the-counter painkiller have anything to do with our assessment of this happiness? What if it is a prescription painkiller? What if it is just a little bit of cocaine? What if it is an unexpected and unlikely-to-be-repeated windfall of money? Because our culture has such strong feelings about cocaine as a happiness source that is not worth the risk, it is illuminating to see the roles the drug has played in other societies, and in our own past.

For most of cocaine's history, it was used only in its leaf form. What people did with the leaves of the coca plant (and still do) is this: you roll up a few leaves and tuck the roll into your cheek. This won't work on its own. You need to use an alkaloid, generally crushed oyster shells or wood ash, to get the drug to work, but these would burn you if you just put them in your mouth. The maneuver, instead, is to take a thin twig dampened with spit, dip it in the ash, and then poke the insides of the leaf-roll so that the leaves break down from inside the packet. Users' spit turns bright green. Sooner or later they feel a buzz, and if they keep a packet of leaves dissolving in one cheek they can walk or work for days with very little food, water, or sleep. This routine has been prevalent for some time: There are forty-five-hundred-year-old statues of people with one bulked-out cheek. People in the mountains of Peru still count time and distance by packets of coca leaves. (One packet will suffice for about a forty-five-minute walk, or about three kilometers.) In the earliest Spanish reports we read of European wonder at the leaves and the sticks (people went about with two pouches slung around their necks, one for the leaves, one for the ash), and at the green spit. When Europeans tried it over here, they felt its delights, and they sang its praises. Yet the coca didn't catch on in Europe: the leaves traveled poorly, the plant was hard to grow in Europe, and the whole thing sounded gross. If you wanted to characterize someone as heathen, making fun of their green spit could only help, and once you had done it, you didn't want green spit yourself. Cocaine stayed out of European use for a long time—and they never did take to chewing it. Or they haven't yet, anyway.

When Europeans started isolating chemical compounds in the very early 1800s, cocaine was one of the last: in 1803 morphine was isolated from opium; quinine and caffeine appeared in 1820; nicotine in 1828. These new plant-derived, active-on-humans substances were all given an "-ine" ending, and when the coca's most active property was isolated in 1860, it received an "-ine" too. Remember that there were thousands of potions in every country that were sworn by for love and for money, and those are two very powerful motivators. Imagine how you would feel if, say, you paid a fortune for a health elixir and did get better; or you gave a potion to your dying mother, and yet she went on dying. Such experiences can make people adamant in either direction—further invested in the idea that the drug was the right thing, or angry. Until scientists proved that there was a chemical ingredient, and gave it a scientific name, how could you know for certain that a drug was the real thing? When the cocaine craze hit, it borrowed nothing from the lab except this legitimacy. Scientists had isolated the chemical, but this knowledge had little to do with the way the cocaine craze started in Europe and the United States.

Corsican Angelo Mariani was living in Paris when he came up with the idea of soaking coca leaves in wine, and he launched his spiked wine from there in 1863. The alcohol broke down the leaves (which could then be removed) and covered their bitter taste. Mariani is famous in the history of advertising because he sent a case to the pope, got a thank-you note, and published it. He did the same with Thomas Edison, Queen Victoria, President McKinley, Sarah Bernhardt, Émile Zola, and two more popes. Big surprise: they all liked the cocaine wine a lot. Most of them were just wild about it; McKinley wrote to say thanks for the case but that he didn't need the introduction as he had been drinking Vin Mariani regularly for a while already. Ulysses S. Grant, Jules Verne, Ibsen, Rodin, all three Lumière brothers, and a hundred others all sent in testimonials. H. G. Wells sent in two drawings of himself, one melancholy, labeled "Before Vin Mariani," and a happy one labeled "After Vin Mariani." Notice there was not a hint of shame or naughtiness in any of this. It was recommended, after all, by presidents, queens, and popes.

Mariani became rich. So did a few enterprising people who copied his idea. One of those was Atlanta druggist John Smith Pemberton. In his fifties, he had already patented a bunch of inventions. He added the kola nut to Mariani's formula and created Pemberton's French Wine Coca. This he marketed as a cure-all, an "intellectual beverage," and a happiness tonic, and it was soon widely adored. When the temperance movement began to gain strength in the United States, Pemberton realized his cocaine-and-wine beverage had an ingredient with a serious PR problem: the wine! There was no obvious way to convert this into a "temperance drink," as wine had been practically the whole drink. Pemberton's nephew tells how Uncle John converted his entire house into a laboratory for Coca-Cola: "an enormous filter made of matched flooring, wide at the top and narrowing to the base . . . was built through the floor of a second story room and the ceiling of the room below." In 1887, the dying Pemberton told a friend, "The only thing I have is Coca-Cola. Coca-Cola some day will be a national drink. I want to keep a third interest in it so that my son will always have a living." That son, Charley, died of a morphine overdose in 1893, at the age of forty. The more drug use there was, the more people died of it. Such stories look like object lessons in hindsight—but most people thought of drugs in a much more open and amorphous way. The cocaine-laden soda was a happiness tonic and seemed to have nothing to do with what killed Charley and so many like him. After a while, though, the connection was hard to ignore. The company began removing the cocaine from Coca-Cola in 1902. Still, it packed a good measure of caffeine, and this, too, was seen as a happiness drug. Slogans for the drink from 1905 include: "Coca-Cola revives and sustains" and "The favorite drink for ladies when thirsty, weary, and despondent." The famous slogan "Coca-Cola lightens worries" showed up in 1907. Again, today, in the West, we refer to caffeine as a wakefulness drug, but many other cultures, even our own a hundred years ago, spoke of it as a happiness drug. Wakeful is productive. Bliss wastes time. The other important work caffeine does is to symbolically erase class. Coca-Cola ads proclaimed, "All classes, ages and sexes drink Coca-Cola." Caffeine's most popular brand name specifically advertised to cross lines of class and gender—just as Lu Yu described teatime as "a moment in the rhythm of the day when prince and peasant share the same thoughts and same happiness."

Oddly, the history of taking cocaine as a distinct drug, based on the modern isolation of the chemical, begins with Sigmund Freud. When it was still a largely unknown drug, Freud wrote to Martha (his fiancée at the time): "In my last serious depression I took cocaine again and a small dose lifted me to the heights in a wonderful fashion. I am just now collecting the literature for a song of praise to this magical substance."[1] Freud felt sure he could come up with an application for the drug that would make him his fortune and allow him to marry. In one of those *hmms* of history, a guy that Freud turned on to cocaine, the ophthalmologist Leopold Konigstein, realized almost immediately that it could be used as a local anesthetic. He became rich; Freud did not. Freud gave the drug to another friend, this time specifically to help kick a morphine habit. This friend liked coke a lot and was dead from it a few years later. So Freud got out of promoting cocaine. But he had enjoyed it a great deal as a happiness tonic, and for all we know he may have continued to enjoy it, to balance those cigars. His study of the drug, *On Coca*, shows his fascination with its effects. I think his experience with cocaine is important evidence about how complex an assessment of cocaine needs to be. We tend to think of drugs as stupidifiers, and we may think so with good reason, but if we give it some thought, we know that drugs are also very interesting.

At the turn of the century, cocaine seemed like a good idea to a lot of people. It worked wonders for those allergic to pollen, who found they could calm their sinuses with a few sniffs, and it was a great relief for toothaches. Also, many advertisers did not shy away from noting that the drug would lift your spirits. From the 1860s to the 1920s there was cocaine in everything from teething remedies for babies, to adult allergy medications, to straightforward happiness tonics. The *Boston Medical Journal* said: "The moderate use of coca is not only wholesome but beneficial." The *New York Times* reported that cocaine had "been applied with success in New York" to cure hay fever and toothache; and the *Therapeutic Gazette* noted the benefits of cocaine and explained, "A harmless remedy for the blues is essential."[2]

When coca use started, and for millennia thereafter, it was mild and, within its area of influence, ubiquitous. It had a lot in common with how caffeine is used in our society. But in the late nineteenth century, coca's contact with Western industrialism, consumerism, science, and transportation led to cocaine growing stronger and more available. At every stage of this development, people set different balances between this form of risk and this form of happiness. As the amount of cocaine in society skyrocketed and the potency of the drug grew in response to market demand, more people got addicted, and more people died. By the 1930s people were campaigning against cocaine use because of the deaths, and also because of the ideological meaning of the drug: it never lost its power as a drug of the lazy or dangerous savage. Coca-Cola took the cocaine out of the drink after a racist "scare" about black men on cocaine raping white women. Another source of racially motivated anti-drug frenzy came from Charles Henry Brent. He was the Episcopalian bishop of the Philippines, though he went there only as ordered— "against my taste and with a repulsion for work in a Latin country." In 1902 Brent set up a moral campaign against drug use, to the great surprise of his flock. In 1910, distressed over his own alcoholic brother, Brent wished all drugs illegal, and moaned, "Society is too selfish, too inconsiderate to give up its pleasant things for the sake of the weak."[3] Brent was just as fired up about "laziness, concubinage and gaming," which he saw as connected to drugs through their "sensuality." His model of righteous severity was widely followed. Drugs started to be seen as dirty—dirty business, dirty bodies.

Others disagreed. The French doctor and detox expert Oscar Jennings wrote in 1909: "It is as a rule assumed that the habitué is necessarily a drug fiend, a cheat and a liar.... That many morphia-takers are narco-maniacs is undeniable, but there are others whose self-control, in restricting themselves to the minimum of morphia necessary to comfort, is infinitely greater than that of an ordinary so-called moderate drinker."[4] Today's "war on drugs" is more right-wing than left, but the temperance movement of the late nineteenth and early twentieth centuries in America was part of the left-wing, progressive movement. Indeed, the ideals of this progressive movement were abolitionism, women's rights, religious freethinking, Indian rights, and temperance. The first big "muck-raking" event, one of the establishing moments of American journalism, was about the drugs contained in patent medicines. Some customers were unaware of the drugs involved. But of course, people know if a remedy makes them feel happy, and when they shop again, they are likely to be knowingly purchasing happiness. People who sold cures with happiness drugs in them were not generally tricking their customers. Everyone understood that what was going on was the selling and buying of happiness drugs. People did not say that, because that is not the way they framed what was going on. But that is what was going on.

As a result of the muckrakers' stories, the government passed the Pure Food and Drug Act of 1906 so that anything with active drugs in it had to be so labeled. Guess what? Drug use went up, not down. In 1907 coca-leaf imports into the United States were twenty times what they were in 1900. Cocaine was both licit and illicit. A Chicago clergyman provides us with a vivid image: one night in 1909 he saw some local boys outside his rectory and the next morning found empty bottles of Gray's Catarrh Powder, each of which delivered eight grains of cocaine.[5] A more aggressive law was championed and carried through by a Dr. Hamilton Kemp Wright; and Wright's reports "raised sensational racist alarms about the cocaine debauches of southern blacks." Wright convinced New York Democrat Francis Harrison to put his name on an act to control and tax opium and coca dealers. In these years, writes historian Davenport-Hines, "it had become almost mandatory to affirm that drug consumers were criminals or degenerates who were enslaved to their habits; that they threatened social order both in cities and the South, impeded business and retarded productivity."[6] As Thomas Blair of the Pennsylvania Bureau of Drug Control wrote in 1919, "the anti-narcotic propagandist has over-stated his case: Virtually there is no opposition, as there is in the matter of [alcohol] prohibition; ... there is inertia on one hand and more or less hysteria on the other. The propagandist has had things largely his own way."[7]

Cocaine passed out of favor for a while. When it showed up in the 1970s, it once again had connotations of productivity and happiness, but by the mid-1980s people once again concluded that the stuff was dangerous, and the use of it began to connote a risky lifestyle, a certain kind of person. As crack, cocaine had a whole new life as part of the underworld, used by people who had become very desperate. Dirty once again. In the present decade, cocaine is back, understood as a risky behavior, but a privileged one. Life is full of risky behaviors, such as mountain climbing or, more commonly, driving in cars, drinking alcohol, having unprotected sex, and being sedentary. Maybe cocaine shortened the lives of the people who carved those ancient statues with the bulked-out cheeks. Still, across coca's history, many people seem to have wanted to live with cocaine in their lives, treating it as a mood-managing happiness drug.

We fall into an error of thinking about drugs in black-and-white terms: that some drugs are good and some are bad. This idea is in the minds of users as well as abstainers. We all inhabit a cultural frame wherein some drugs are for normal people and some are for naughty people—and none of them are understood as happiness drugs. The only drugs our present-day culture might call happiness drugs are modern prescription psychopharmaceuticals—and even these are more likely to be called antidepression than pro-happiness. In fact, these drugs can provide happiness, just not usually euphoria, and it is strange that we do not seem to want to acknowledge that. Good drugs are only supposed to cure illness, to fix something wrong, so if we want a drug to be seen as a good drug, we don't talk about it as providing happiness. Obviously, it is not reasonable to claim that our modern antidepressants are curative, as against all the other drugs in history, which were recreational, deviant, or dulling. Today, drugs that are illegal are also not credited with providing happiness, though people privately know that they do, and some will snicker when they have to disavow the joy of drugs. Illegal drugs, including cocaine, are usually depicted in the culture as mentally and emotionally numbing, perhaps providing a kind of sped-up (in the case of cocaine) or slowed-down stupid giddiness and blank disassociation. Sometimes illegal drugs are culturally connected with euphoria, but never happiness. But, again, many people have chosen cocaine as a happiness drug for millennia. Only in the last two centuries has it become common to use cocaine in such strong forms, though. Perhaps if North America and Europe had simply adopted (and made commercially viable) the leaf-chewing habit of the South Americans, we would today find cocaine at newsstands and supermarkets, like chocolate and coffee. Likewise, imagine that coffee beans could be cultivated so that they packed more of a euphoric punch. In a hundred years coffee might be illegal, and high-school kids will hide behind the gym to chew the beans. Will they look back and be amazed that once upon a time it was served on almost every street in America? (Or maybe it will be chocolate that gets chemically goosed into providing a serious high, and thus becoming dangerous.) People love having a "cure for the blues" or, to use Freud's words, to be "lifted … to the heights in a wonderful fashion." It is very hard to determine what harms and what benefits are really attached to a drug in any given form, in any given amounts, over any given amount of time, as taken by any given person. So we are largely thrown back on culture, which is also tricky.

The cultural meaning of every historically important drug is complicated by its origins, but cocaine may be particularly burdened. Trenchant images of race, class, and sex are attached and reattached to different forms of the drug. A crack high is drawn as brief euphoria amid despair and a coke high as a brightening up of preexisting abundance. The middle class may use cocaine in order to declare access to luxury, but it doesn't mind if celebrities are punished for their use of it, precisely because it is a mark of their privilege. As for crack, the middle class uses it as a definition of falling out of their class and into desperation. They may use it anyway, perhaps almost ironically. If one drug has been culturally associated with criminal violence, it is crack, and again, as in times past, control of cocaine has been envisioned as controlling certain racial groups and controlling the poor. This influences our ideas about the drug. With a clean cultural slate, perhaps we would look to ways to make milder forms of the drug and would make these available. For happiness. A lot of people would do it sometimes. Perhaps we should see today's Wellbutrin use as similar to the use of coca leaves in ancient Peru.

The other major historical happiness drug is opium, the juice from the unripe seedpod of the flower *Papaver somniferum*. Why is opium a happiness drug? One of the purposes of this discussion is to show that although we think of opium as a drug of melancholy stupor, this is not how most of history experienced the drug. If you have done any historical research in old newspapers, you have seen ads for laudanum. The women in the ads are happy and lively, and the list of things the potion claimed to cure was long. Laudanum was opium and wine, spiced with saffron and cinnamon, and it certainly made a person feel better. You also may have seen movies where women drink from little vials in their purses and get catatonic. What's up with the widow character on the series *Deadwood* who clings to a tiny vial? Laudanum. What were doctors thinking? Was it similar to what doctors think today when they prescribe Prozac? Why does it seem so irresponsible? My short answer to those questions is that people thought that poppy-juice happiness was good happiness. Part of the reason for this was the same situation as with cocaine: when modern labs, industrialism, and markets got their hands on the drug, superstrong versions were developed. In this case, heroin was one of the strongest and most dangerous, and the resulting deaths inspired some unusually virulent rejections of poppy drugs in our culture. The other reason nineteenth-century doctors handed out opium was because people of that time needed its medicinal properties more than we do. Where cocaine is great for allergies and toothaches, opium has a more important medicinal punch: it stops coughs and diarrhea. During the nineteenth-century epidemics of tuberculosis and dysentery—due in part to overcrowded industrial cities—the ability to steady the lungs and bowels was an incontestable blessing. For many people, it was health and happiness in one bottle.

Opium use for happiness goes back to prehistory. Sumerian ideograms dated to 4000 B.C.E. proclaim the poppy to be a plant of joy.[8] In Egypt, around 1500 B.C.E., the *Therapeutic Papyrus of Thebes* prescribed opium as a cure for babies suffering colic. It worked, I presume, whether you gave it to the mother or the babe. It is in the fifth century B.C.E. that we see poppy use first discouraged, by the great philosophic author Diagoras of Melos, and caution was also encouraged by the West's first great doctor, Hippocrates. For the next five hundred years we hear of it as a pleasure of the elite, mostly, but in the time of the great ancient doctor Galen (130–201 C.E.), its use seems to have become quite general. Galen wrote that there were lots of recipes for theriac, but that if it doesn't have poppy juice, it is no good.[9] Theriac was known first as an antidote to poisons from animal bites, and later as a cure for disease. The basic recipe was opium and spices in honey. Galen recorded his preparations of theriac with scientific detail. He kept especially careful records of his theriac preparations and their effects on his most famous patient, the philosopher-emperor Marcus Aurelius.[10] Depending on whether the emperor wanted to get happy, kill pain, or get to sleep, Galen adjusted the opium content of the honey potion. Galen's records show that Aurelius was able to assess the quality of the ingredients and the dosage of a day's potion, and he knew to abstain when need be for matters of state.

This is amazing and important. Marcus Aurelius is a great hero to almost everyone who has spent any time with him. He wrote the emperor of all self-help works, *Meditations*, a book packed with stunning insights laid out as poetic exercises—meditations for more perspective on oneself, others, and reality itself. He was also a famously wise ruler. Master historian Edward Gibbon (1737–1794) called the reign of Marcus Aurelius "the period in the history of the world during which the condition of the human race was most happy."[11] Aurelius held the vast Roman world, but he knew that what we really want, we want from the people close to us, and that beyond a certain point they won't give it to us—because they have their own thing going on. We have to figure out how to do the best we can under these circumstances. As we saw, his advice is to not try to be like Plato and imagine a new utopian world, but rather to try in moderate earnest to make this world good. If you pay meditative attention to the vast size of time and the span of the universe, you will be more realistic, more generous, and will develop a better sense of humor about yourself. Remember that either life means something, or it means nothing and is all chance; either way, you need to relax, get through your panic, and get in on the game. Most of all, if you can remind yourself, in stable and unstable times, that change is the nature of things, you will be wise and able to help others. Only don't think that if you are perfectly generous, kind, and wise you will escape the derision of humanity: people have something nasty to say about everyone. If you are perfect, Aurelius warned, they will still be a little glad when you are dead and no longer around making them aware of their not being as perfect as you.

Life is hard for everyone sober, even Aurelius! The only person ever known as a great emperor and true philosopher was on a steady, carefully managed diet of opium highs. Apparently, there are limitations to wisdom, and happiness is often managed by a variety of means. Borrowing his *Meditations*, and not his opium, is perhaps like borrowing the idea of a bicycle but building yours with only one wheel (perhaps to cut down the chances of your getting a flat tire). Of course I am not actually advocating that university classes in Roman philosophy come with a lab fee for the dope, as we know that these drugs hold bad trouble for anyone who cannot control their use. Still, it is possible to see modern psychotropic drugs, or booze, or various smokable plants, in similar terms. Also, modern prescription opiates like Vicodin are used by people to feel happy, to take away pain, and to help them fall asleep. Marcus Aurelius tinkered with his chemistry just as we do today. Later, Galen was personal physician to the Emperors Commodus and Severus. Commodus refused theriac. Severus demanded the exact preparation given to Marcus Aurelius; he so admired the already legendary emperor that he did his drugs. As for Galen, throughout his practice he continued to recommend the use of opium as a cure for headaches, asthma, coughs, colic, fevers, and melancholy. People would henceforth refer to mixed opium drugs not only as theriacs, but now as *galenes* or *galenics.*

In the Middle Ages the word *theriac* became *treacle,* and Venice became the European source of it. The Venetian treacle recipe was a thick, sugary syrup with opium in it, and was used medicinally for every ill. When opium began to be regulated, the syrup without the drug was sold as a cheap sweetener, still called treacle in Europe, but usually called molasses in the United States. In the medieval European period, while opium use was slowly changing from ancient Greek theriac to Renaissance Viennese treacle, most of the new writing was produced in the Muslim world. One of Muhammad's sacred tenets was that Muslims are not supposed to drink alcohol. That doesn't mean they never did, or don't now, but it was and is a strong prohibition. Since they should not drink, Muslims took a great deal of hashish and opium. (It reminds us of the Mormons disallowing caffeine while sipping ephedra.) There were apothecaries in the Arab world in the eighth century, and they stocked a huge variety of opium medicines. Some great Muslim champions of opium were the famed ninth-century doctors Yaqub ibn Ishaq al-Kindi, Sabur ibn Sahl, and Muhammad ibn Zakariyya al-Razi. Al-Kindi wrote in praise of opium medicine; Sabur's four-hundred-odd opium or poppy recipes are for a range of remedies. Included are those that are "good for old age" and one that, if given to a healthy person, "will protect him from all pains and diseases." The recipe in which he called for the most opium was for a drug that he says is "useful for stupidity and lethargy, and sharpens the brain." One recipe for a drug to combat depression, charmingly called "Food for Sorrows," included henbane and opium. If you have seen the HBO series *Rome,* you may have noted that after Titus Pullo was much injured in his unsuccessful execution, his doctor gave him henbane to keep him asleep, comfortable, and immobilized. The drug was also used to get high.

In the late ninth century, the great doctor and rationalist philosopher Abu Bakr al-Razi was the first to introduce the use of alcohol for medical purposes. Al-Razi was surprisingly secularist and more concerned with people's well-being than with religious rules. In surgery, alcohol could alleviate pain and give a patient courage, so he used it. The great medieval doctor and philosopher Avicenna was apparently quite a hashhead. In the vast Islamic world, opium use by the masses was well established by 1000 C.E. Hashish was very popular, too. European visitors have been amazed, throughout history, by the sheer quantity of these substances commonly consumed in the Islamic world. Sufis seem to have had a rich history with hashish and the use of hallucinogenic plants like mandrake and thorn apple.

There were opium apothecaries in Europe in the twelfth century. In 1240, the German emperor Frederick II issued an edict separating the practices of medicine and pharmacy, creating professional pharmacy and providing government supervision over pharmacists, who were to prepare drugs "reliably, according to skilled art, and in a uniform, suitable quality." The pharmacy, or apothecary shop, was decorated with curiosities:

giant stuffed crocodiles, turtle shells, the horns of rare beasts, and other treasures. As time went on, objects symbolic of magic and natural curiosities were joined by increasingly fanciful glass vials for the various opium concoctions. Drugs in artful majolica and faience jars were soon on display in the better homes as well. The apothecary shop became a meeting place where one could buy drugs, take drugs, and sit around gazing at the stunning glass sculpture, the taxidermy, and the exotic scientific devices. Medical consultations were available there, but this was also very much about socializing. These were places to relax and talk.[12] Many doctors and pharmacists of the seventeenth century based their whole curative system on administering various opium potions.

The early modern period has a widely documented record of happiness-drug use in every stage of life. Scottish physician John Brown decided that emptying people out with bleedings and purgatives was a bad idea. His *Elementa medicinae* of 1780 claimed instead that the sick should be stimulated, and no stimulant was better than opium. He relied on it himself as a cure for gout. A contemporary of Brown's, the physician George Young, took laudanum for his cough. He described a typical case from his practice, a woman with "a disponding mind" (a nice old term for depression) who "received more benefit from opium alone than I could well believe: it not only suspended her menstrual flooding, but all her fears and gloomy ideas." This part raises eyebrows: "All her friends advised her to lay aside the use of opium, lest it should by habit become necessary, but she whispered privately, that she would rather lay aside her friends." She reported that she stopped taking opium after a few months of her pregnancy and then just kept it on hand in case of "disponding fits."[13] Samuel Johnson and Benjamin Franklin both took opium in their painful final years, because, to use Johnson's word, dying makes one "gloomy" and opium helps. Baron de Montesquieu praised Asian and Muslim medicine for "seeking as carefully for remedies for sadness as for the most dangerous diseases."[14]

America turned away from bleeding and toward drugs in response to the medical treatment of George Washington in 1799. Washington was ailing, and his doctors bled him. Four times. He got weaker and weaker, and he died. Today's physicians studying the case agree that those bleedings killed Washington, and people at the time thought so, too. It wasn't as if this was the first time someone had died from a bleeding, but this was George Washington. As with Rock Hudson's revelation of his having AIDS, when a larger-than-life figure experiences something, it can affect us in a larger-than-life way. People did not want to be bled to death, so they stopped going to regular doctors. Also, as Americans, it flattered our sense of independence to be able to reject the old European bleeding techniques and offer some homegrown American cures. America rejected established medicine in favor of the idea of allowing the body to heal itself, assisted by drugs, spa treatments, and special diets. Morphine use rose, and had a less alarming fatality rate than the bleeding that had preceded it.

Thomas De Quincey's 1821 *Confessions of an English Opium-Eater* was perhaps the first tell-all of its type: the man needed money, didn't mind attention, and sold his story. He had had facial neuralgia at about nineteen years of age and was given opium for it, and liked it; he said he spent one night a week on laudanum at a concert or opera. He felt troubled or anxious a lot, and distraught about his scolding mother, so he took the drug to feel good. He also reported private drug experiences that were sometimes happy, sometimes ecstatic, and sometimes bad or even horrible—sometimes all at once: "I was the idol; I was the priest; I was worshipped; I was sacrificed."[15] Despite his ambivalence, for over a century this book was credited with drawing people to opium. They wanted the heightened experience, and the comfort. Elizabeth Barrett Browning recorded her experience with drugs: "Opium—opium—night after night! & some nights, during east winds, even opium won't do."[16] It's a droll line. Samuel Taylor Coleridge credited his marvelous poem "Kubla Khan" to a morphine dream. His daughter, Sara, wrote more directly to the drug. Consider a poem she wrote called "Poppies," which speaks for itself. Here are most of its stanzas. (Note: *Cramoisy* is an old term for crimson cloth. Herbert is her young son.)

> The Poppies blooming all around
> My Herbert loves to see,
> Some pearly white, some dark as night,
> Some red as cramoisy;
>
> He loves their colors fresh and fine
> As fair as fair can be,

But little does my darling know
How good they are to me.

O how should thou with beaming brow
With eye and cheek so bright
Know ought of that blossom's power,
Or sorrows of the night!

Oh then my sweet, my happy boy,
Will thank the poppy flower,
Which brings the sleep to dear ma-ma
At midnight's darksome hour.[17]

Sara had had some hard times. As Virginia Woolf put it in a biography of Sara, "Children were born and children died," some very young. Sara herself died of breast cancer at forty-eight. Herbert Coleridge, for his part, was considered a genius in his time. As a young philologist, he set the foundation for what became the Oxford English Dictionary. Herbert hadn't much time to see it grow, though: he was dead of tuberculosis at thirty-one. It was 1861. For those in pain, the idea that relief is only one sip or one needle prick away is perhaps a little dreadful, but mostly wonderful. One had friends who did it, too, and one had heard of people who had lived to old age, sipping and pricking all the while. And who could deny the wishes of a woman or a man who had made so many trips to the graveyard?

Drugs were respectable and publicly relied upon in ways that would shock people today, and opiates were openly adored. Listen to the fabulous character Lydia Gwilt in Wilkie Collins's 1866 novel *Armadale*:

Who was the man who invented laudanum? I thank him from the bottom of my heart whoever he was. If all the miserable wretches in pain of body and mind, whose comforter he has been, could meet together to sing his praises, what a chorus it would be! I have had six delicious hours of oblivion; I have woke up with my mind composed; I have written a perfect little letter … ; I have drunk my nice cup of tea, with a real relish of it; I have dawdled over my morning toilet with an exquisite sense of relief—and all through the modest little bottle of Drops, which I see on my bedroom chimney-piece at this moment. "Drops," you are a darling! If I love nothing else, I love you.[18]

Increasingly, from 1850 to 1915, middle- and upper-class women got so comfortable with their habit that at the opera they took syringes out of little beaded opium purses and shot up in front of everyone, the way you might take a mint. Blouse fashions changed to hide the track marks. As drug historian Barbara Hodgson put it, "[Their] habit was no longer confined to stuffy sickrooms; it was out in the open, even flaunted. Reliable statistics are difficult to come by, but numerous references imply that morphine syringes among the theatre crowd were as common as cigarettes."[19] That's a lot of morphine. In 1908 the San Francisco *Examiner* featured a large article on jeweled morphine paraphernalia: syringes and vials. In a similar magazine one could find ads for "Mrs. Winslow's Soothing Syrup" (opium and 90 proof alcohol), with text that advised mothers that the medicine "should always be used" for teething and colic. The writer Alphonse Daudet described a socialite addicted to morphine in one of his novels: "There's a whole society like that.... When they get together, each of these women, carrying their little silver cases with the needle, the poison ... and wham! In the arm, in the leg.... It doesn't make you sleep ... but one feels good."[20] Imports of opium to Britain in 1839 totaled 41,000 pounds, and by 1852 they reached 114,000 pounds a year. In 1840 the United States imported about 24,000 pounds of opium a year; by 1867 the number hit 135,000 pounds, and by the 1890s, over 500,000 pounds came in annually.

In William James's classic *The Varieties of Religious Experience* (1902), he tells us about the "New Thought" cure (more on that in the next chapter), and cites a woman's letter to him. I like how casually her letter mentions drug use, and how clearly such a thing represented cure rather than recreation or self-indulgence.

Life seemed difficult to me at one time. I was always breaking down, and had several attacks of what is called nervous prostration, with terrible insomnia, being on the verge of insanity; besides having many other troubles, especially of the digestive organs. I had been sent away from home in charge of doctors, had taken all the narcotics, stopped all work, been fed up, and in fact knew all the doctors within reach. But I never recovered permanently till this New Thought took possession of me.

Hear that? The way she mentions narcotics? It is the sound of historical difference: a great age of medical confidence in narcotics was over, but our period of narcotic criminality was still a long way away. Across all these periods, some people took drugs for happiness.

Where did all the legal opium go? I think one important answer is that when TB was tamed, the need for a cough suppressant and painkiller diminished. The bacteriological revolution's triumph over so many fatal childhood illnesses removed some of the strongest reasons for dosing oneself or one's family. Also, some addictions used to be indistinguishable from common wasting illnesses, and once these latter were eradicated, the junkies really stood out: there was no crowd of blameless ghouls among whom an addict could blend. Another big answer is that legal opium or close copies of it are still around in a great variety of our painkillers—codeine and morphine, of course, and Vicodin, Percocet, and Demerol, for instance. The philosopher Bertrand Russell, in his 1930 work *The Conquest of Happiness*, wrote, "I am not prepared to say that drugs can play no good part in life whatsoever. There are moments, for example, when an opiate will be prescribed by a wise physician, and I think these moments more frequent than prohibitionists suppose."[21] The acceptance or rejection of drugs on the basis of cultural bias—that is, on the basis of historical ignorance—is a cognitive trap. The way out, of course, is a self-induced shift in perspective, an opening to the notion that the way we see things is not objective and true, which means we have a responsibility to try to undermine this trap, to think otherwise. The years go slowly if they are not happy. We have drugs that can help make people happy—short-term bliss, long-term grins. It is not fair to you or to the many people you influence to dismiss this question without giving it some serious thought.

7

Religion and Revelation

Growing up in the second half of the twentieth century, in New York, I gleaned from my culture the message that drugs and religion were very much interconnected, but not for us. In my family, the holidays were our time for wine: four cups at Passover. We lived on Catholic Long Island, and wine was clearly important to that religion, too. Wine was the blood of Christ. Yet neither Manischewitz nor Gallo was an *entheogen*, a drug that brings religious experience. Was wine at Passover or at Communion *ever* an entheogen? I search the Hebrew and the Christian Bibles for references to wine, and I do not see anything that looks even vaguely entheogenic. Wine is discussed as a food staple in the Hebrew Bible, and people are warned against excessive drinking in the Christian Bible, but throughout both, wine is a happiness potion and people thank God for it, over and over. In Judges an explanatory clause is nicely revealing: "wine, which cheers both gods and men." Psalm 104 says wine "gladdens the heart of man."[1] When Jesus turned water into wine, it was "the first of his miraculous signs.... He thus revealed his glory" and won his disciples' faith (John 2:11); later it was wine that he chose as the matter of his blood. Alcohol cheers, gladdens, and allows for the miraculous and the glorious to be revealed. All this shows that wine has long served as a happiness drug in the Judeo-Christian world, but it does not suggest that people were inducing drunken religious revelations. I had heard of entheogens only as part of Native American religion.

Native Americans seemed to know what to do with teenagers. Their culture provided challenges; meaning; episodes of freedom; festivals with drugs, music, campfires, mad dancing; and adults on hand as guides. Native Americans took peyote and became one with hawks or wolves, and when they came back from their trips, other people wanted to know what they saw. Peyote trips were counted as part of the relevant universe. As I was growing up, this appeared to seem sublime to my elders, and it certainly sounded sublime to me. Today the culture has preserved interest in these religious drug experiences, though both the legal mood and the general sensibility have grown less awesomely respectful. There may have been bad trips among the Native Americans. There may have been sad people who found escape in peyote and took it too often. There must have been fatal bad batches, or simple overdoses. There is little evidence of these, though. Peyote seems to be a drug that brought common happiness by occasional use: an incredible experience now and again that made the rest of life seem less limited. Something transcendent was part of real life. It was shared across generations in a way that felt vital. Drugs, by definition, can engender euphoria, feelings of insight, love, and freedom. Religions have used drugs throughout history. Private individuals who take drugs have also used religion to help manage the emotional and psychological meanings of their psychotropic experiences. William James wrote that while some mystical states of consciousness are dismissed by public opinion, we find witnesses to these states in "private practice and certain lyric strains of poetry."

> I refer to the consciousness produced by intoxicants and anesthetics, especially by alcohol. The sway of alcohol over mankind is unquestionably due to its power to stimulate the mystical faculties of human nature, usually crushed to earth by the cold facts and dry criticisms of the sober hour. Sobriety diminishes, discriminates and says no; drunkenness expands, unites, and says yes. It is in fact the great exciter of the Yes function in man.... It makes him for the moment one with truth. Not through mere perversity do men run after it.

For the uneducated, says James, drink "stands in the place of symphony concerts and of literature," and he also says that it is a great mystery that something so magnificent is available to so many of us "only in the fleeting earlier phases of what in its totality is so degrading a poisoning."[2] He's not missing that it is degrading, and a poison; he's praising it *anyway*, as so many people do by their actions. Modern rhetoric is completely missing this part of the conversation.

James added that nitrous oxide and ether also stimulate a mystical consciousness.

> Depth beyond depth of truth seems revealed to the inhaler. This truth fades out, however, or escapes, at the moment of coming to; and if any words remain over in which it seemed to clothe itself, they prove to be the veriest nonsense. Nevertheless, the sense of a profound meaning having been there persists; and I know more than one person who is persuaded that in the nitrous oxide trance we have a genuine metaphysical revelation.
>
> Some years ago I myself made some observations on this aspect of nitrous oxide intoxication, and reported them in print. One conclusion was forced upon my mind at that time, and my impression of its truth has ever since remained unshaken. It is that our normal waking consciousness, rational consciousness as we call it, is but one special type of consciousness, whilst all about it, parted from it by the filmiest of screens, there lie potential forms of consciousness entirely different. We may go through life without suspecting their existence; but apply the requisite stimulus, and at a touch they are there in all their completeness.... No account of the universe in its totality can be final which leaves these other forms of consciousness quite disregarded.[3]

Philosophical insights about reality, James continued, suggest a secret that most of us do not usually have any sensory access to: "Those who have ears to hear, let them hear; to me the living sense of its reality only comes in the artificial mystic state of mind."

Aldous Huxley found mescaline helpful. His *Doors of Perception* (1954) took its title from a William Blake line: "If the doors of perception were cleansed, everything would appear to man as it really is, infinite."

Just in case you are already thinking *hippy*: Huxley, a lithe man, small of shoulder, wore suits, glasses with thick circular lenses, and kept his dark hair swept neatly back from his forehead. He explained that scientists had noted a chemical similarity between mescaline and adrenaline (epinephrine), which made psychiatrists wonder if we were not all capable of producing the chemical. Mescaline-trip descriptions were very similar to the way schizophrenics described their reality. The question became "Are mental disorders due to chemical disorders?" Psychologists had begun trying the drug on themselves to see what their schizophrenic patients perhaps saw. Huxley decided to try it, too, in the spring of 1953.

His experience led him to believe a notion he credited to Cambridge philosopher C. D. Broad, but which we can see proposed earlier by Schopenhauer and others. In Huxley's words: "The function of the brain and nervous system is to protect us from being overwhelmed and confused by this mass of largely useless and irrelevant knowledge, by shutting out most of what we should otherwise perceive or remember at any moment, and leaving only that very small and special selection which is likely to be practically useful." We thus have access to a flood of perception, even total perception, "but," Huxley explained, "in so far as we are animals, our business is at all costs to survive. To make biological survival possible, Mind at Large has to be funneled through the reducing valve of the brain and nervous system. What comes out at the other end is a measly trickle, but the kind of consciousness which will help us to stay alive on the surface of this particular planet."[4] On top of this atrocious narrowing of reality, humanity invented languages, which are not only systems of speech, but also hidden philosophical systems, and assumptions about what constitutes similarity and difference. Huxley's claim was that workaday consciousness is a slim and selective vision of reality. Peyote gave you a world so different, you couldn't believe your brain could put on a show like that. It becomes clear that this brain of yours is putting on a show all the time. If you experience a complex and entirely convincing hallucination—say, your horse talking to you in the language of the Plains Cree—you know something new about perception and about the world.

Take the drug, and "all kinds of biologically useless things start to happen."[5] Huxley listened to the recordings of his conversation under the influence of the drug, and unlike James reported, "I cannot discover that I was then any stupider than I am at ordinary times." Instead "the eye recovers some of the perceptual innocence of childhood"; also fun was that interest in space and time falls almost to zero. A drawback to the drug was that "the mescaline taker sees no reason for doing anything in particular." That was a problem in part because the trip lasted a bit too long for our culture. (We like a party to last one night.) Charmingly, he put the conundrum in terms established by Jesus: "Mescaline opens up the way of Mary, but shuts the door on that of Martha. It gives access to contemplation—but to a contemplation that is incompatible with action and even with the will to action, the very thought of action."[6] If you haven't read your Christian Bible lately, Mary wants the bliss of truth, and Martha wants to be a good hostess and get everyone lunch:

> As Jesus and his disciples were on their way, he came to a village where a woman named Martha opened her home to him. She had a sister called Mary, who sat at the Lord's feet listening to what he said. But Martha was distracted by all the preparations that had to be made. She came to him and asked, "Lord, don't you care that my sister has left me to do the work by myself? Tell her to help me!"
>
> "Martha, Martha," the Lord answered, "you are worried and upset about many things, but only one thing is needed. Mary has chosen what is better, and it will not be taken away from her." (Luke 10:38–42)

Huxley said we couldn't just be Marys, seeking truth all the time, that we have got to be Marthas, too, for the sake of civilization. It is a telling mid-twentieth-century line, as if Jesus were so obviously wrong here that he should be casually corrected, like a silly schoolchild. The Kingdom sounds terrific, but we've got to set the table or nobody eats. Let us at least consider that Jesus was right here, and that contemplation of the true reality of existence might sometimes be allowed to take precedence over household chores.

It was scary to face a "reality greater than a mind," but this fear is part of what excited Huxley: "The literature of religious experience abounds in references to the pains and terrors overwhelming those who have come, too suddenly, face to face with some manifestation of the *Mysterium tremendum*." Amazingly, he announces:

> All I am suggesting is that the mescaline experience is what Catholic theologians call "a gratuitous grace," not necessary to salvation but potentially helpful and to be accepted thankfully, if made available. To be shaken out of the ruts of ordinary perception, to be shown for a few timeless hours the outer and the inner world, not as they appear to an animal obsessed with survival or to a human being obsessed with words and notions, but as they are apprehended, directly and unconditionally, by Mind at Large—this is an experience of inestimable value to everyone and especially to the intellectual.[7]

People will probably always need what he called "artificial paradises," he mused, because "most men and women lead lives at the worst so painful, at the best so monotonous, poor and limited that the urge to escape ... is and has always been one of the principal appetites of the soul."

Huxley borrowed H. G. Wells's phrase "doors in the wall" to refer to "art and religion, carnivals and saturnalia, dancing and listening to oratory." Huxley noted that all these activities require so much effort and collaboration that they are not for daily use: "For private, for everyday use there have always been chemical intoxicants. All the vegetable sedatives and narcotics, all the euphorics that grow on trees, the hallucinogens that ripen in berries or can be squeezed from roots ... have been used by human beings from time immemorial. And to these natural modifiers of consciousness modern science has added its quota of synthetics — chloral, for example, and benzedrine, the bromides and the barbiturates." Sighed Huxley, "Countless persons desire self-transcendence and would be glad to find it in church. But, alas, 'the hungry sheep look up and are not fed.'"[8] They go to church, he explained, but it isn't a place of transcendence. For solace they turn to alcohol: "God may still be acknowledged; but He is God only on the verbal level.... The sole religious experience is that state of uninhibited and belligerent euphoria which follows the ingestion of the third cocktail. We see, then, that Christianity and alcohol do not and cannot mix. Christianity and mescaline seem to be much more compatible." Huxley ends the book with these remarkable words: "The man who comes back through the Door in the Wall will never be quite the same as the man who went out." He will be "wiser but less cocksure, happier but less self-satisfied," better equipped to understand the relationship of words to things, and "the unfathomable Mystery."[9]

Walter Pahnke's famous 1963 "Good Friday Experiment" was conducted as research for his Doctor of Divinity degree at Harvard. Pahnke invited twenty divinity school students to take part: ten received a capsule with some psychedelic mushrooms in it, and the other ten students got a placebo. A Good Friday church service was held for them while psychiatrists, doctors, and scholars watched and later held interviews. The psychologist and famous proponent of psychedelic drugs Timothy Leary was among the experts on hand. Pahnke devised a test for a "mysticism scale," and the stoned students scored much more mystical than the sober ones. Here are the self-rated experimental and control groups, shown as percentages of highest possible scores at six-month and long-term follow-up.[10] Imagine each subject being asked how much he felt sacredness, for instance, that day.

The Good Friday Experiment

	Experimentals		Controls	
	Six-Month	Long-Term	Six-Month	Long-Term
Unity—Internal	60	77	5	5
Unity—External	39	51	1	6
Transcendence of time and space	78	73	7	9
Deeply felt positive mood	54	56	23	21
Sacredness	58	68	25	29
Objectivity and reality	71	82	18	24
Paradoxicality	34	48	3	4
Alleged ineffability	77	71	15	3

A follow-up study in the 1990s found that, for most of the participants in the study who got the drug, the experiment was remembered with gratitude. Many had become clergymen. One described the experiment thus: "Religious ideas that were interesting intellectually before, took on a whole different dimension. Now they were connected to something much deeper than belief and theory."[11] Interestingly, this man served on Stanford's campus as a drug counselor after finishing divinity school and later became pastor of a Unitarian congregation in Florida, and a husband and father. Years later, after much experience in the world and in the church, he is still enthusiastically positive about his day of drugged insight and revelation. Note also that for one student, the drug experience had been awful—full of paranoia and feelings of isolation.

The big news at the time of the Good Friday Experiment was that even our modern religion could be enhanced through drugs. People were moved enough to wonder if drug experiences accounted for the origins of religion. In 1963 Mary Barnard asked readers of the *American Scholar*, "Which ... was more likely to happen first, the spontaneously generated idea of an afterlife in which the disembodied soul, liberated from the restrictions of time and space, experiences eternal bliss, or the accidental discovery of hallucinogenic plants that give a sense of euphoria, dislocate the center of consciousness, and distort time and space?"[12] Other well-known thinkers have also argued that drugs invented religion.[13] A year later the great historian of religion Huston Smith published an article in the *Journal of Philosophy* entitled "Do Drugs Have Religious Import?"[14] The piece had a huge impact—and no wonder, as it remains insightful today. At one point, Smith asks the reader to choose which of the reported enlightenment experiences was drug induced:

I) Suddenly I burst into a vast, new, indescribably wonderful universe. Although I am writing this over a year later, the thrill of the surprise and amazement, the awesomeness of the revelation, the engulfment in an overwhelming feeling-wave of gratitude and blessed wonderment, are ... fresh.... The knowledge which has infused and affected every aspect of my life came instantaneously and with such complete force of certainty that it was impossible, then or since, to doubt its validity.

II) All at once, without warning of any kind, I found myself wrapped in a flame-colored cloud.... there came upon me a sense of exultation, of immense joyousness ... [then] an intellectual illumination impossible to describe.... I did not merely come to believe, but I saw that the universe is ... a living Presence; I became conscious in myself of eternal life ... that the happiness of ... all is in the long run absolutely certain.[15]

Along with the fun of this game of Guess Who Was High (the first one), I cite the quotations here because they seem packed with information. Most important, the first quote is further evidence that the positive effects of the drug experience last long past the event. Also important is the quotations' vivid illustration of Smith's point—not just that drug and religious enlightenment are similar, but that they are so similar and so amazing that they essentially prove each other. The normal life of us bony bags of goo often feels lonely and meaningless, but every once in a while we are treated to a different experience that is exactly as real and all encompassing as this one, but that feels united, sublime, and blissfully happy. Smith reminded his readers of a variety of religious uses of drugs, including "the wine used in our own communion services," peyote, soma, marijuana of the Zoroastrians, the benzoin of Southeast Asian Zen, the *pituri* of the Australian aborigines, and "the mystic kykeon that was eaten and drunk at the climactic close of the sixth day of the Eleusinian mysteries."[16]

What are the rules about how we use alcohol, painkillers, depression medication, illegal drugs, and religion? Since the culture uses all these drugs anyway, is it not reasonable to look at the way these drugs function for us? People take Percocet to go to church, and they have a great time. A religious flock that wouldn't come drunk and doesn't even know much about hallucinogens might well be drugged with all sorts of mood stabilizers and pain relievers and hence find it easier to have a religious experience. Government stores in India today sell bhang cookies: what looks like a rum ball will take about two hours to kick in, and then you find yourself in a world without time, a world of unfathomable instability, for about six hours. We talk about ourselves as pill-popping, but maybe people in most of history have taken a lot more drugs than we do, and maybe the particulars of these drugs have something to do with the particulars of given cultures. More important than their differences, though, are the similarities between drugs. They all make it clear that our common perceptions are just one turn on the dial. Drugs show that your brain is capable of the absolute reshaping of the information your senses bring it; and that leads to the revelation that your brain is always shaping the character of what your eyes tell it. But I'm not talking about only visual hallucinations: say, all your hairbrushes acting like cancan dancers. During the high, drugs make psychological symptoms disappear. They can reverse lifelong personality traits. The frightened get brave, the uptight come loose, the clown turns contemplative. Even simple booze can allow revelatory experiences about love, purpose, friendship, and hope. Also, mystical experience has been considered good for people's happiness. I had thought that these experiences were brought on by putting the mind and body through tough exercises. Now I am sure that at least some mystical experiences were brought on by ingesting or smoking something.

Today, the politics are very silencing, and that may be just as well, because drugs can be very dangerous. Still, the relation of drugs to mystical experience is worth thinking about. Let's see if we can pass over "the war on drugs" of the last decades, and read Smith's conclusion with fresh eyes:

> The conclusion to which evidence currently points would seem to be that chemicals can aid the religious life, but only where set within a context of faith (meaning by this the conviction that what they disclose is true) and discipline (meaning diligent exercise of the will in the attempt to work out the implications of the disclosures for the living of life in the every day, common-sense world).
>
> Nowhere today in Western civilization are these two conditions jointly fulfilled. Churches lack faith in the sense just mentioned, hipsters lack discipline. This might lead us to forget about the drugs, were it not for one fact: the distinctive religious emotion and the one drugs unquestionably can occasion—Otto's *mysterium tremendum, majestas, mysterium fascinans*.[17]

The Otto mentioned here is the much-beloved religious thinker Rudolph Otto. His *Idea of the Holy* (1923) is still held up as the modern work that best understands the ecstasy of trembling in awe at the great mystery of God. Smith wants to experience awesome trembling; this is all very important to him. I love the phrase "hipsters lack discipline," and the fact that what churches are found lacking is here termed "faith." Smith saw that transcendence outside a context of disciplined effort might not be wonderful. Still, for many people, a common life punctuated by some euphoric drug experiences is a life made fuller, more tremendous. This sort of drug use might be just as effective for overall happiness as the regular use of a drug that brings us to a level mood.

There are also "highs," flights of exquisite happiness, that come to a person out of nowhere, unaided by drugs. These can also last an hour but inform a lifetime. You never hear anyone talking about this. By contrast, we all know that sometimes a black mood hits us, relatively unprovoked. It is a commonly mentioned phenomenon: sometimes we all feel dark and a little hopeless. So it is surprising that the culture does not have a similar message about happiness. If you hear of a man living in luxury or success, you may console yourself by remembering that he has sad moods, like everyone. But when things are bad, it is rare to hear anyone say, "Well, even if nothing ever helps, stay alive, because you will experience some inexplicable waves of happiness, some of them so intense as to be revelatory." The only place you hear talk of this is in poetry. Consider Yeats's evocation of sudden, transcendent happiness in this section of the poem "Vacillations":

> My fiftieth year had come and gone,
> I sat, a solitary man,
> In a crowded London shop,
> An open book and empty cup
> On the marble table-top.
> While on the shop and street I gazed
> My body of a sudden blazed;
> And twenty minutes more or less
> It seemed, so great my happiness,
> That I was blessèd and could bless.

Such a moment is usually so hard to articulate that we tell only a few people about it, ever, and draw on it for our sense of wonder across decades. Epicurus said that the only positive attribute we can ascribe to the gods is that they are always happy like this. It was a deep recognition of the holiness of this elusive, ghostlike thing, the mood of bliss. The detail Yeats gives seems to be there to show us that the setting was ordinary, had nothing to do with what happened to him. But since it would be nice to be able to re-create his elation, let's look anyway. I am moved to see the crowd, and what I assume to be an empty cup of tea.

Poetry seems to be the only place these flights of joy are described. Jane Kenyon's poem says that our blues defend our blues, but that when happiness shows up, the blues have no chance; happiness will win:

Happiness
There's just no accounting for happiness,
or the way it turns up like a prodigal
who comes back to the dust at your feet
having squandered a fortune far away.

And how can you not forgive?
You make a feast in honor of what
was lost, and take from its place the finest
garment, which you saved for an occasion
you could not imagine, and you weep night and day
to know that you were not abandoned,
that happiness saved its most extreme form
for you alone.

No, happiness is the uncle you never
knew about, who flies a single-engine plane
onto the grassy landing strip, hitchhikes
into town, and inquires at every door
until he finds you asleep midafternoon
as you so often are during the unmerciful
hours of your despair.

It comes to the monk in his cell.
It comes to the woman sweeping the street
with a birch broom, to the child
whose mother has passed out from drink.

It comes to the lover, to the dog chewing
a sock, to the pusher, to the basket maker,
and to the clerk stacking cans of carrots
in the night.

It even comes to the boulder
in the perpetual shade of pine barrens,
to rain falling on the open sea,
to the wineglass, weary of holding wine.

In such heaviness, out of such dark clouds, happiness is the magical uncle who comes out of the sky in a plane just for you. People get happiness in occasional flare-ups, in all sorts of situations, despite ignorance and despite wisdom. The presence of wine does not escape us.

Consider this final poem of happiness oddly descending. It is by Raymond Carver, more famous for his short stories than his poetry, and more famous for his drinking and desperation than for his later, happier years.

Happiness
So early it's still almost dark out.
I'm near the window with coffee,
and the usual early morning stuff
that passes for thought.

When I see the boy and his friend
walking up the road
to deliver the newspaper.

They wear caps and sweaters,
and one boy has a bag over his shoulder.
They are so happy
they aren't saying anything, these boys.

I think if they could, they would take
each other's arm.

It's early in the morning,
and they are doing this thing together.

They come on, slowly.
The sky is taking on light,
though the moon still hangs pale over the water.

Such beauty that for a minute
death and ambition, even love,
doesn't enter into this.

Happiness. It comes on
unexpectedly. And goes beyond, really,
any early morning talk about it.

It is only for a minute that death and ambition do not enter into it. Still, you can feel his minute. Happiness comes on unexpectedly. There was also a cup of coffee. I'm being a little cheeky, mentioning these beverages, but I think it is fun that even in these master expressions of unprovoked happiness, we still find tea, wine, and coffee.

Imagine that three hundred years into the future our culture has become interested only in the metaphysical, as has occurred in some extraordinary and long periods of history around the world. There have been huge Buddhist societies where most young men were sent to meditate for a decade or two, because the spiritual cause of the country was paramount and the society believed that everyone was in it together; this is also true for Judaism and Catholicism. Imagine our world morphing into such a place: without a myth system—just a huge respect for anything that takes you out of the common, mundane world. With science to oversee quality and distribution, and advise us on use, we would, of course, want to make use of these amazing substances that we have found on the planet. It would be crazy to ignore (and even disdain!) the fact that our bodies have a capacity for responding to a chemical, such that the entire world changes and we see something distinctly other and distinctly good.

8

Drugs Today
Music and Solace

Where do we expect to smell marijuana smoke in public in our culture? Soon after the drug theorists of the mid–twentieth century
lamented that there was no place in the culture for reverential tripping,
a powerful youth music culture came into being and provided such a
venue. Music culture is a mass market now, but it still operates as a
place for transcendence for some. Some of the participants of the music
and drug culture want transcendence and are taking the one offering of
it that is available. If the music is beloved, the setting is dramatic, the
crowd is passionate, the addition of some alcohol or some drugs can tilt
the experience into bliss—a sense of being at one with the world. In
Native American religion, music and drugs were similarly mixed, but
the participant could count on guides in whom there is real and merited
trust; social rules for interaction with strangers; audience participation in
music and dancing; and wide social praise for seeking transcendence, for
carrying on the culture's tradition. At a concert or in a club you are lucky
to get a few of these. At a concert, loyalty to the musician or to the type of
music played bonds the crowd and provides a sense that a tradition is
being upheld. At clubs there isn't an audience: everyone is in it. At both of
these places it is okay to let people see you show emotion, move your body
rhythmically, sing, make faces, act a little out of control.

  Think about how science today tries to get to happiness. It is big on
drugs. But not the kind of drugs that are related to listening to live music

or dancing (or, for that matter, seeking mystical revelation)—even though so many people, especially young people, say they are never happier than when they are a little high and listening to live music, or dancing. This is a trance of value. Every era has trances of value; we cannot live without them. But when they walk us into the same wall, over and over, we can try to snap out of them enough to fix the problem. The acceptance of curative happiness drugs and the rejection of "party" happiness drugs is based on a degradation of "party" happiness. It isn't reasonable or at all scientific. Also not reasonable or scientific are the drugs we allow and disallow for solace.

Today a drug commonly associated with a transcendent experience is Ecstasy, or MDMA. Ecstasy induces a feeling of intense empathy, sensuality, and contentment. The effect is profound enough that people who have tried Ecstasy report having a new idea about all existence and perception. They also report that the conviction remains with them long after the experience. Interestingly, Ecstasy works in the same way Prozac does, by inhibiting the reuptake of serotonin in the brain, but Ecstasy also causes a surge of serotonin to be produced, so it works much faster, and the effect is more intense. It can have a pretty hard letdown period: some report feeling low the day after, or longer. Ecstasy also has a lot in common with the amphetamine Adderall, which is increasingly replacing Ritalin as a daily drug to help children calm down and concentrate in school. It is notable that the patent on Ecstasy has expired, so no pharmaceutical company stands to gain huge windfalls from marketing it if were it to be made legal. MDMA, like LSD, was originally used by psychiatrists, in this case as a tool for therapy. Ecstasy was developed in 1914 and was widely used in psychotherapy in Europe in the 1970s because it helped patients to relax and talk. In Germany it was regularly recommended by marriage counselors. What makes MDMA bad and Prozac okay? A well-known 1996 article in the *Economist* put the question as such:

> Every week, according to the most conservative estimates, half a million Britons take a pill to make them happy. This pill was originally developed as an appetite suppressor. Now it is an adjunct to partying. In America, some 5m people regularly take a different sort of pill. This one was developed as an anti-depressant....

The British users are breaking their country's law. The Americans are not. Which raises an important question. If it is not acceptable to take [MDMA, or Ecstasy] ... to make you feel happy when you just want to have fun, why is it acceptable to take [Prozac] ... to make you feel happy if you are not actually clinically depressed?[1]

The article suggests that the answer is a "pharmacological Calvinism" that causes people to reject any drugs that are used by some people for fun. Most of us assume that drugs are a less authentic way of attaining psychological goals, but there is no reasonable argument for this. The *Economist* cited studies by Eric Hollander, of New York's Mount Sinai School of Medicine, showing that treating obsessive-compulsive disorder with drugs changes the patients' brains (disentangling the action of four groups of nerves) in the same way as treating them with psychotherapy. The article concluded, "It hardly seems that one method is morally inferior just because it is easier."

Julie Holland, an attending psychiatrist at Bellevue and on the faculty at the New York University School of Medicine, edited a compendium of opinions about MDMA, and in it many doctors and scientists talk about the drug's potential for creating happiness and alleviating symptoms.[2] Dr. Andrew Weil, the famed health theorist, attests to the therapeutic uses of MDMA, writing that "MDMA can give you a chance to have a new perspective on your body, and ... that's part of breaking old habits."[3] More specifically, he has seen people physically relax, for the first time, on MDMA. That, Weil says, has meant long-lasting reduction in allergies, back pain, and digestive problems. Also, the memory of the immense optimism brought on by the drug has lasting effects. Weil also argues that an MDMA experience can make a person dying of cancer feel a lot better about dying, as well as significantly reducing their physical pain, and that the drug belongs in the regular arsenal of palliative care.

Other authors have collected stories not from scientists but from spiritual leaders (who seem to be less willing to give their full names). A Catholic monk, Brother Bartholomew, says he has taken MDMA twenty-five times in the last ten years. "While using MDMA, he has experienced a very deep comprehension of divine compassion. He has never lost the clarity of this insight, and it remains as a reservoir on which he can draw," reports author Nicholas Saunders.[4] A rabbi at a London synagogue spoke to Saunders of the value of the drug for terminal patients because it allowed them "the feeling of oneness." The rabbi added that "taking drugs is like reaching the top of a mountain by cable car instead of through the toil of climbing—it can be seen as cheating but it gets you to the same place."[5] A Rinzai Zen monk named Bertrand, in his seventies, agrees: "The result is in every way as real because it is the same."[6] The consolation of a bit of euphoria and a palpable sense of the unity of life seems worth offering. Should we hold back this opportunity for insight into spiritual peace? It does not seem justifiable.

Whatever kind of consolation MDMA should be allowed to be, in the 1980s and 1990s it was a drug for the club and rave scene. Particularly in England and across Europe, and later in the United States, the Ecstasy dance scene was big. There was often a sense of communal revelation in the drug, the music, and the dancing. It is predominantly young people, and they have the sense of being in on something new and important, benevolent, and honest.

Consider another perspective. The man being interviewed on *Oprah*, Jim MacLaren, tells us that he used to be a football player but while on his motorcycle he was hit by a New York City bus and lost his leg. At first his life seemed over, but he refound life, retrained himself as a runner with a high-tech new leg, and became a champion in his new sport. Then one day he was riding his bike on what was supposed to be a closed track and a van plowed into him. This time he was informed at the hospital that he would never again feel or move from the neck down. He doubted he could refind life yet again, but, from his stair-climbing, cool-looking, high-tech wheelchair, he beamed at the audience that life had indeed refound him, that he has worked hard and regained some marked limb movement, and that his purpose now is to spread a message of motivation and optimism. His second accident was a further blessing; he is at last at true peace. Oprah asks, Did you really need the second accident to get there? *Oh yes*, he says. *I needed to sit down.*

The reason I tell you this story at this juncture is that Oprah asked him to speak to the period after he was released from the hospital, paralyzed, and before the time he re-embraced life. Yes, he responded, with a nod of responsible contrition, when the second accident happened, depressed and suicidal, he left home and spent nine months in Hawaii on a cocaine binge. Hawaii and cocaine, he explained, was what he liked as far as locations and drugs went, and he stayed with it until he stopped wanting to, which happened when he found himself talking to people who were not there. Scared and tired of it, he came back to the mainland and cleaned up. As his story was framed here, a man of amazing inner resources has admitted to a period of failure, disowned it, and returned his narrative to one of strength and gratitude for life. A more integrated story could be told from this same drama, wherein a man hit so hard by life that he might not have ever recovered found that he was able to use drugs to attain a period of rest, unreality, and euphoria. During this time he was able to come to terms with his new situation, and invoke, once again, his amazing inner resources. The drugs were no error; they allowed the man's salvation.

Doctors confronted by a patient in profound, understandable sorrow used to give the patient opium, and as our historical doctors tell us, some of these patients limited their use of the drugs to what made life bearable. A lot of people nursed mild, pleasurable addictions. It was considered good for you. If a despondent woman—say, her husband was shot dead in front of her and died in her arms—refused to take a happiness narcotic, something that gives you a burst of euphoria, her doctor would strongly suggest she change her mind and take the drug. If she was unable to care for her child under this treatment, relatives could be asked to take the child in until the mother regained her strength. If the woman would not relent, the doctor would explain that her misery would break down her health and that if she wanted to be there for her daughter later in life, now she needed rest and respite. Then she would be weaned off the drugs. Remember the woman whom William James quoted, the one who tried all the narcotics, and all the diets, but nothing really worked for her until she embraced the "mind cure"? Presumably she liked or needed the narcotics for a while, and then stopped.

Creatures from outer space who could see the inhabitants of Earth but not hear us would see similar behavior from the people of both centuries. However, if all they could do was hear the people of both centuries, they would judge the cultures to be completely different. This splitting the view between sight and sound doesn't bear much further inspection, but it is an evocative notion here. All I am trying to say is: See how people act? That's how people act. Then we pour thick blankets of culture on the whole thing, full of extreme value judgments that determine people's lives, and yet are as changeable as ocean waves, rolling over and over. The woman whose husband was killed and died in her arms was not a creature of my imagination. I was thinking of a woman named Ashley Smith. As you may recall, in 2005 Smith came to fame when she was taken hostage in her home by an escaped convict who had killed a random four people that day, and she was able to convince him not to kill her, and even to let her go. She told him about her own jail experiences, about the death of her husband, about her little girl of whom she had lost custody because of her recent devotion to methamphetamines. She gave the guy some meth, which he had never had before and which calmed him down. She herself was so moved by her desire to see her daughter again that she decided then and there to quit her use of the drug. The big thing was that she convinced him that there was life after this day of horror: he could go to prison and pay for what he had done; he could find God, and then bring his wisdom to others in prison. So he didn't kill her or himself, or anyone else, and he didn't risk getting shot on the run. He stayed put and let her leave to go see her daughter, at which point she called the cops, who sped over and collected him, peaceably. That was some pretty good humanism on her part. Americans responded with wonder and acclaim for her. But only when her book came out did we find out about the drugs. It is so degraded a behavior that she had to apologize to her reader before she could even start, naming her book *Unlikely Angel*. All I want you to do is imagine how all these actions would look if the whole thing happened in the nineteenth century. Some drug use helps get people through hard times, and it might be better if the culture had room for this, and if this kind of drug use were something one discussed with one's physician.

In general, Americans have always been worried about unproductive happiness. After all, capitalism and democracy are sustained by unsatisfied desire and alert participation. What if everyone felt super, just rolling out of bed, scaring up some food, and then relaxing? But the concern is not only that contentment might harm capitalism and democracy; we have been worried about contentment as its own problem. Indeed, through much of the twentieth century we insisted that the reason democracy and capitalism were better than communism was precisely because communism might produce too much contentment, and thus turn us into drones. In the midst of the Cold War, the film *Invasion of the Body Snatchers* drew on this fear, showing vital, frantic Americans replaced by relaxed, dully-happy alien versions of themselves. In reality, you have to be exceptionally miserable to choose a dull, drugged, uninspired stasis. It is terrible to note that many people who take drugs do so because they are, indeed, exceptionally miserable, or at least powerless. The despair of poverty and racism leads people to want to be numb, and sexism does, too: women in Iraq take massive quantities of Valium, which is available without a prescription and costs about two cents a tablet. When drugs ease a tough, unfair experience, they also may allow it to continue, and it is often not obvious whether a drug to ease the pain is a blessed mercy or an insidious tool of the oppressors. If you feel bad and you *can* change your real-life situation, you still might take some palliative potions, but you certainly would not want that to interfere with your ability to flourish.

We are not happy when we are too sluggish or too wired, when our loved ones feel neglected, when we get nothing done, when we know we are hurting our health, when we can't remember what happened. If you can take a drug and be happy and still not run into any of those problems, that is probably a good definition of "not taking too much." Human happiness is always going to be generated through problem solving in our projects and in our relationships. The frustration of trying is part of what happiness requires, and the agony of real growth is necessary to a good life. Here are some of the things long-term happiness requires in the short term: studying for exams; caring for children who are going through unpleasant phases; being responsible at work; forgiving friends and spouses who have hurt you terribly; keeping the promises of marriage; maintaining your home; going to the doctor and dentist; saving money; finding something nice to eat; taking a walk; visiting your extended family. Drugs can provide true euphoria, and they can provide good-day happiness. They cannot provide the goods of good-life happiness. Good-life happiness absolutely requires putting in a variety of tiring efforts, many of which are better done sober. The rewards are not merely the result of the struggle; they *are* the struggle, seen from a different angle, from a different vantage point in time.

So, no pill can touch overall good-life happiness, because that is an intellectual and emotional assessment, not a bodily mood. When people worry over what would become of us if we had a true happiness pill, they are asking about good-day happiness. The reason I dismiss the question as it might pertain to euphoria is that there are euphoria happiness pills already, and most of humanity does not sustain interest in them. For the great majority of people who try them, nonaddictive euphoria drugs (psychedelic mushrooms, for example) do not become a lifelong or frequent habit. Euphoria lacks the anchors of our common existence. That is part of what makes it euphoria. But the absence of these anchors can be scary and keeps you from your common doings and pleasures. Also, of course, these drugs are illegal. And, since they are illegal, people worry about the lack of safety controls. But even if they were legal and relatively safe, it seems likely that euphoric hallucinogenic drug use would be rare for most people. So what is left to worry about is good-day happiness pills. Modern psychopharmaceuticals are not supposed to make you joyous or even happy. Doses start low, and when the subject reports that his or her symptoms have abated, that is the dose that is maintained. Why do we treat drug euphoria as a problematic side effect? We do not even have a word that would indicate a high level of regular happiness. I'll use *rapture*. We fear that rapture leads to addiction. Addiction can be unpleasant. It can also be pleasant, of course (you probably enjoy needing coffee), but when it is bad, it is awful. Also, people fear that rapture would diminish our willpower to do the chores required for life happiness (though it does not always have this effect). I agree that both points are a matter of concern, but the dismissal of rapture in response to these two threats seems extreme.

Shouldn't we be nicer to ourselves? Drugs are a historically consistent part of how we manage the emotions of being a mammal who can think, and remember, and ask why the world is not otherwise. Even if you are healthy and shrewd, the world can break your heart. Woody Allen and Schopenhauer both memorably lamented that nature was essentially "one enormous restaurant," plants eating other plants, big fish eating little fish, indeed, everyone hounding everyone else, offering "only momentary gratification, fleeting pleasure conditioned by wants, much and long suffering, constant struggle, *bellum omnium*, everything a hunter and everything hunted, pressure, want, need and anxiety, shrieking and howling; and this goes on … until once again the crust of the planet breaks."[7] It is hard to be such a clever mammal that we notice this. Of course you need a drink. Or something.

The legal antidepressants that we use today will change meaning. I don't know whether they will prove unhealthful, or will be found to extend life and be put in the water for everyone; whether feminists will reject them or demand them; whether children raised on them will someday denounce today's doctors or thank them. We may feel we missed all the fun of original cocaine-laden Coca-Cola, but someday people may speak dreamily of the good old days when doctors handed out prescriptions for SSRIs as if angels were handing out Toll House cookies. Or people will buy them at the supermarket and wonder what the fuss was. Two hundred years from now, what will be the history of caffeine? Of nicotine? Of cold medicines? Studies are beginning to show that depression and anxiety have a significant adverse effect on lifespan and health. In the future, we may become more aware of the health ravages of depression, and find that happiness drugs have a less adverse effect. What if, because of some disease or comet, momentary happiness becomes a lot more important than long life, for some people or for all of us? We would have to come up with a new calculus of happiness risk. That idea calls our attention to the fact that we now use some such calculus.

Psychic pain is similar to physical pain. A broken leg that was never set properly may hurt throughout one's whole life. You need to have that leg looked at, and probably broken and reset. For psychological pain, we often need to engage in years of talk therapy. In the meanwhile, drugs to lessen the pain make sense. Drugs to lessen psychic pain cannot be sorted out into piles of curative versus masking. Modern antidepressants are not curative medicine fixing a broken you. There is nothing wrong with your little gray brain blobs, labeled A and B as in the television commercial. I repeat: there is no evidence that depression is caused by a deficiency of serotonin. Your psyche got knocked around. You never fixed it, and you're still in pain. You want to feel better. As for euphoria, it seems that many people through history, many religious people among them, have made use of drugs for a blast of euphoria every once in a while, and have been able to use such experiences to enliven their everyday lives over years of abstinence. Hallucinogens like Ecstasy, mescaline, and peyote are euphoric, but they also change your reality, and for some people, the use of these drugs even once in a lifetime brings insight, comfort, and spiritualism.

The lesson of history is that drugs are mostly good. They are in our world, like food and sunshine, and we have always made use of them. In our world there is a shocking abundance of everything, so it is hard to know what's for you, and when to stop—and this goes for drugs as much as for anything else. We manage this deluge by all sorts of classifying systems about what is good and what is bad, what is healthy and what is unhealthy, what makes us seem weak and what makes us seem strong. Taking drugs can do you harm, but so can not taking drugs. Many physicians and researchers argue that people heal faster when they are not in physical distress, such that painkillers can be seen as part of a healing treatment. Depression, anxiety, loneliness, and grief shorten people's lives in a variety of ways, from simple failure to take care of oneself, to suppression of the immune system, all the way to suicide.

It is clear to me that adults who want to know more about happiness ought to employ drugs in that effort. Many people try drugs, let's say out of curiosity, and do not find the experience rewarding. If such people do not long to affect their happiness, and their curiosity on the subject is low, I would not counsel them to press on and try more drugs. But here are some circumstances in which to try a potion:

- If you are tortured or even bothered by heavy moods.
- If you long for a break from your "symptoms" or merely from your personality.
- If people keep telling you that you should ask your doctor for an antidepressant.
- If you want to know more about the nature of reality and how the mind creates it.
- If you want to set up conditions for a mystic's revelation.
- If you want to have an intense communing experience with someone.
- If you want to have a good time on a given evening.
- If you want to dance and be social, but you are too inhibited.

I would not counsel the use of illegal drugs for happiness, because despite the moral call to civil disobedience in relation to unreasonable laws, if you get caught, you won't be happy. That leaves room for trying a lot of drugs, especially if you are willing to travel. But let us begin right here in the United States and see what we can do. We can drink tea and coffee not only as productivity drugs, but also as happiness drugs. That might mean only a change in one's mind, but it may involve switching beverages, and maybe finding different places and ways to drink them. We can refuse to let cultural ideas stop us from trying pharmaceuticals that might otherwise interest us. We can take it as a responsibility to experience something of this nature. When we drink alcohol we can think about it as a possibility for minor metaphysical events, not only as a technique to numb ourselves; we can see it as a different kind of intelligence rather than as stupidity.

Smart people have been analyzing the meaning, benefits, and drawbacks of drugs for a long time. Centuries. Millennia. Their overall message is that human chemistry and the chemistry of the world are separated by a porous border. In a normal life we are naturally affected by all sorts of external chemicals. The question is not whether to take drugs, but which ones, and with what expectations. Modern pundits who claim we are dosing ourselves too freely are not speaking with historical knowledge or philosophical sense. Keeping your mind in only one place is not a very assertive way to relate to life, to search it for happiness, or for truth. The point is to be equally robust in analyzing and controlling your experience. And being savaged by inner pain doesn't do anyone any good. Through history, many people have let their drug use get out of hand, but many people have managed quite well. It is worth thinking about. In this section I have tried to jostle your sense of categories when it comes to drugs, partially just because jostling categories allows us to get a fresh perspective on reality. That is to say, the issue of drug categories has been here used in the service of a wider investigation into how people think and how thinking changes over time. The trances of value seem dependant on categories, so you can wake yourself up by confronting the fact that the categories are fantasies. You are not ever going to be happy all the time. Since we are human and the universe is not, there will always be disappointment. Sour charm is in no danger of being cured. But as all graceful-life philosophies agree, sometimes it is good to let yourself be helped and let yourself be happy, and if you block off whole categories of solace as forbidden, you might miss out on what you need.

Money

Everybody knows that money doesn't buy happiness. Yet it sure feels like it would, if only we had a little more. These opposing convictions are both wrong. They are the two key features of our modern myth about money. The truth revealed by historical research is this: money can buy happiness, and it already did. Most Americans are living in a blissful world free of the sorrows, sores, and losses experienced by most people throughout history and throughout the world. What we do with money nowadays is to remind ourselves of this triumph. Above poverty, it is not very important how much money you have. What counts is that you spend what you have in a way that makes sense to your feelings and to your reason. To do this, you need historical context. The world is not so disenchanted and rational as people think; our culture has fixations that drag us around. When we are more aware of them, we can see how money works for us, and we can augment our pleasure with understanding. We need to remember that most people through history have been racked by work that was bloody-knuckled drudgery, the periodic desperate hunger of their children, and for all but the wealthiest, the additional threat of violent animals. Nowadays a lot of what we use money for is a symbolic acting-out of these triumphs.

It is a modern myth that money and happiness are unrelated for the wise and in direct proportion for the shallow. They are never unrelated, and, above the poverty line, they are never in simple direct proportion.

Why do the wise—religious or philosophical—all say that money doesn't buy happiness and that you should give away all your stuff and "toil not," like the lilies of the field? The answer is that they really don't all say this. Only a few do, and they do not offer it as a way of living in the world. They offer it as a way of leaving the regular world in favor of an outrageous new community. Diogenes recommended that you walk away from all your property and live naturally with his group, like a pack of friendly street dogs. Jesus said to divest yourself of all your worldly possessions and go with him to embrace the coming kingdom of heaven. Eastern and Western mystics also hold that owning anything at all is a serious roadblock to enlightenment. But certainly in Western civilization, and in some of Eastern civilization as well, a lot of ardent followers live at a distance from this stark advice, choosing to miss out on the promised ecstatic otherworldly happiness and instead keep their stuff and their stake in this one. The same goes for philosophers and artists: some tell you to leave the world in order to more deeply reach states of beauty and awe, but most philosophers and artists want love and common pleasures and seek their beauty and awe from within the mundane world. In the great wisdom literature of history, this more average path—staying in the common world—is often considered the best one.

Consider one example from the history of philosophy and one from the history of religion. Aristotle said that if you are searching for a happy man, look for a man who is materially at ease. His *Nicomachean Ethics* is primarily devoted to the relationship between virtue and happiness, but it also acknowledges that wealth, fame, friends, and honor help flesh out a good life. Happiness, Aristotle held, requires a degree of comfort, and an ability to support those who are naturally dependent on us. Money gives us the means to entertain guests, do favors for people when appropriate, avoid debt, and negotiate the political marketplace. Likewise, from the camp of religion comes Koheleth—the author of Ecclesiastes—who advises us that if you have perfumed oil, you should anoint yourself. Dress well, he says; have a good time. True, he notes that riches can do you more harm than good: "The sleep of a laboring man is sweet, whether he eat little or much: but the abundance of the rich will not suffer him to sleep." But still: "Behold that which I have seen: it is good and comely for one to eat and to drink, and to enjoy the good of all his labor that he taketh under the sun all the days of his life, which God giveth him: for it is his portion." If you have "riches and wealth, and ... power to eat thereof," Koheleth says, it is your role to "take [your] portion," and to "rejoice in your labor." In other words, as he famously put it: "Eat, drink, and be merry." For Aristotle, and for Ecclesiastes' Koheleth, riches could be of use in happiness. Still, money was of very small concern. The big issue was always more about getting one's mind right.

The wise generally say that if you have money, you should enjoy it, and they argue that money is of use for a good life. However, sages and philosophers almost never say it is worth it to go out and get rich as a primary life project. Aristotle and Koheleth both brilliantly caution us away from seeking happiness by seeking money. To prove my point briefly and iconically, I will let one of capitalism's greatest proponents caution us as to the folly of that plan. In 1759, a while before his *Wealth of Nations* of 1776, Adam Smith made his name with a work called *The Theory of Moral Sentiments*. We think of Smith as capitalism's original and optimistic theorist, but he was wise and worried, too. In this earlier book, Smith says that when a poor man's son has ambition, it is a curse. The condition of the rich "appears in his fancy like the life of some superior rank of beings," and to reach it, the young man "sacrifices a real tranquility that is at all times in his power."[1] If he attains wealth, "he will find [it] to be in no respect preferable to that humble security and contentment which he had abandoned for it." Power and riches are high-maintenance machines "contrived to produce a few trifling conveniences to the body." The machines "must be kept in order with the most anxious attention, and ... in spite of all our care are ready every moment to burst into pieces, and to crush in their ruins their unfortunate possessor.... They leave him always as much, and sometimes more exposed than before, to anxiety, to fear, and to sorrow; to diseases, to danger, and to death." Of those without wealth Smith writes: "In what constitutes the real happiness of human life, they are in no respect inferior to those who would seem so much above them. In ease of body and peace of mind, all the different ranks of life are nearly upon a level, and the beggar, who suns himself by the side of the highway, possesses that security which kings are fighting for."

The fact that the beggar is sunning himself tells us that we are to think of Diogenes, who is almost always pictured in recline, and who answered Alexander the Great's offer of any favor with the request that the young conqueror get out of his sun. And Alexander did wish he had "the security" of Diogenes, as Smith would say two thousand years later. But what if you are not Diogenes but a person with a modest home and three kids and are in danger of losing your job? Is that really just as conducive to happiness as is wealth? Shakespeare was always mentioning the stress and anxiety of being rich and powerful, but he has to work hard to convince the rest of us, and perhaps himself, too. In his plays, kings sigh that they are under so much pressure that they cannot get any sleep and that they envy their poorest subjects for this reason. But the rest of his characters are sure harried by their own problems. One of the things people want money for is so they can worry less. It seems right to me: what could be worse than insecurity about paying rent, maintaining health insurance, and paying down debt? But for decades now, economists and sociologists have been finding that money does not reduce worry. A 1976 study of worry by Frank M. Andrews and Stephen B. Withey found that, above the poverty line, "[t]here are virtually no differences associated with socioeconomic status."[2] Researchers have looked at the issue in a number of ways and yet find it very difficult to confirm the common assumption that wealth helps to cut down on worry. Another study, in 1981, showed that people with less money and less education worry about their health and income, whereas those with more money and more education worry about their spouses and children.[3] Alas, this seems right, too. If you are up all night worrying about what your son is doing, who can tell you that worrying over the mortgage is worse? We can worry about two things at once, but researchers keep finding that the general amount that worry impinges on one's life is not in any kind of direct relationship with how much money you have. It suggests that above poverty level, worry is best approached through wisdom, not conditions. Money must be good for something else.

9

Happily Ever After

Money pays for a lot of surveys conceived to discover the results of American material bounty—that is, money. What the studies show has struck many people as profoundly paradoxical. Since World War II, our average standard of living has gone up enormously: a chart for average household luxuries escalates like a stairway to heaven. A lot of families have two cars, a washer-dryer, several televisions, telephones, refrigerators, vacations, new clothes, and, occasionally, new furniture. Meanwhile, a chart showing happiness flatlines or even declines. Asked the same questions that had been asked Americans in the 1950s, people in the 2000s report themselves to be no happier. The paradoxical ratio has been confirmed by a great variety of research. Political scientist Robert E. Lane has provided us with an elegant compilation of these studies in his *The Loss of Happiness in Market Democracies*.[1] Over and over researchers have discovered that abundance does not correlate with happiness. This seems especially strange because the abundance has been guided, at every stage, by desire. The whole idea of democracy and capitalism was that they were based on giving the people what they wanted, inviting them to vote their preferences and supplying their demands.

People with money want more money. Does that mean the money we already have doesn't work, so we always think we need more? Or that the money we already have is working great, so, of course, we want more? We know that people want more. The standard way of judging the well-being of a nation is its GDP, or gross domestic product—the total market value of all goods and services produced in a country in a given year. That of the United States is high, of course. Yet a recent survey showed that less than half of the population could agree that they have enough money to lead the kind of lives they want to.[2] Another study asked whether people would like more money, and an overwhelming majority said yes. We keep associating more money with more happiness even though research studies and popular wisdom (from inspirational posters to clichés of philosophy class) keep telling us that money does not bring happiness. Is there something wrong with the thinking of everyone who has money but wants more, or is there something wrong with the research studies and the popular wisdom? Both. Our searching after money is in part animated by a fallacy that the powerful and lasting happiness created when money lifts us from poverty can be repeated as money lifts us yet higher. But money still occasions a lot of opportunities for short-term happiness and for an overall sense of life's drama and progress. The research studies mostly back up the popular wisdom, and the popular wisdom is substantially wrong. By contrast, the uncommon wisdom of most great philosophers includes a respectful valuation of money. Furthermore, nowadays money serves to replace a lot that money stole. In important ways, the idea that money cannot, today, buy happiness is a remarkable myth. Of course money doesn't always bring happiness, but it has the capacity to bring some. People are sometimes naive about how tricky it is to get the good out of money, but that doesn't mean they are wrong to suppose that there is some good to be gotten out of it.

There is a clear correlation between abject poverty and unhappiness. When an increase in money means the difference between uncertain and unpleasant food and shelter on the one hand and stable and decent food and shelter on the other, the conclusion that money brings happiness is inarguable. After this first jump, though—from hungry and cold to full bellied and warm—the story of money is a story of sharply diminishing returns. Food, housing, and possessions that become nicer and nicer add to our happiness less and less. What I call the *abundance inference* is the mistake of thinking that first step out of penury is representative of further advances. The inference seems supported because all of us, poor or rich, get a burst of joy upon achieving or receiving a new delight—scientists these days talk about it as a strong dopamine surge. Research has shown that the effect of rising out of poverty is lasting: five and even ten years later you still self-report more happiness than you did when you were without basic resources.

Why do people report themselves to be as happy as they were back when we all had less? Well, for one thing, we are comparing two societies that are both majestically wealthy in comparison to almost all societies throughout history. Neither the surveyed Americans of the 1950s nor those of the 2000s were struggling with endemic distress— hunger, pain, humiliation. And average people who are not in such distress are statistically more likely to call themselves happy than not. Whatever happy means to us, we have set it to correspond with our normal state. But happiness, in each of its three forms, requires maintenance. For a good day, you have to have something a little extra to look forward to, and you cannot spend too much of the day in tasks that are drudgery to you. For euphoria, you have to take risks and usually suffer a little. For a happy life, you have to see some changes that feel like progress. You can get all this without money, but nowadays it is remarkable how effective money can be in helping get it. Given the choice to do this happiness work without money or with money, the people seem to have voted. They find it easier to do the work *with* money.

Giving oneself a midday treat by drawing on inner resources is an unbelievable drag in comparison to buying something nice to eat or doing a little window-shopping with an option to buy. And, of course, you are going to need to use your inner resources, too, to see such acts not as an empty and meaningless exchange but as the kind of thing that makes us complex apes happy: a little give-and-take, a little novelty, a little confirmation of self-value. Money can help in the service of lifelong happiness, too. Showing yourself that you are grown up enough to be a parent is reasonably well accomplished by purchasing a large, safe car and a home full of furniture. People get emotional work done with money. Other people get emotional work done without money. All of them report that they feel sort of happy. Of course, with or without money, some people don't do their happiness maintenance work, and they feel a little lousy. Happiness maintenance work is creating things to look forward to on a daily basis; arranging some peak experiences for yourself occasionally; and making sure the overall story of your life has some feeling of progress and growth.

One guess about the happiness flatline since the fifties is that the statistics reflect a case of rising expectations and competition: keeping up with the Joneses. You might become dissatisfied with your lovely backyard lawn when your friends all put in swimming pools. Philip Brickman and Donald Campbell named this the "hedonic treadmill" some four decades ago.[3] The idea is that in this consumerist, competitive world, whatever you purchase or achieve makes you happy for only a moment, exciting your appetite for more rather than satisfying you. When it comes to promotions, film roles, or a new handbag, one makes you want two. But is this really about wanting more? Sometime in the hours after you have eaten lunch, you will begin to be aware of a desire for dinner. Is this an *eating treadmill*? It would be bad if we could not enjoy breakfast because we already wanted lunch. Maybe you enjoy the lawn, and then later put in a deck, and enjoy that, too, and then a pool, and you enjoy that, too. If you thought money gestures were going to be permanent, this is going to look like a treadmill. Otherwise, maybe not.

Money can also make you feel like you are part of a group. But you don't have to have much money for this; you just have to have the same amount as the group you are in—so long as you are above the line of real poverty. Consider an argument I'll call *falling down with the Joneses*. In the British film *Hope and Glory*, set in London during the Blitz, there is a scene where two female friends rummage through racks of used clothing at a "make-do-and-mend session." Molly says, "God, how I hate all this scrimping and squalor," and Grace responds: "I don't mind it. It was harder before the war. Trying to keep up appearances. Now it's patriotic to be poor." For some people, some of the time, community status is more important than the quality of one's possessions. Note, however, that Molly said she hated the squalor; she did not say she hated no longer being envied. Walking around a shiny-floored department store, touching lovely pristine blouses, is an objectively different experience from walking around a dingy room with your coat on and buttoned against the chill, looking at worn, tired clothes. Likewise, chances are good that you genuinely enjoy gliding around your new in-ground pool on a hot summer day. That is to say, envy and rising expectations do not explain everything. There are things in life that can make you happy, in a combination of innate mammalian pleasure on the one hand and, on the other, symbolically important phenomena that work like hypnotic triggers for happiness: palm trees and bubble baths, for instance. Some of this you can buy.

All else being equal, people will generally guess that they are doing well. A 1996 study showed that two-thirds of people think they have above-average earnings.[4] Economist Robert Lane explains it like this: "[I]t seems to me that the market ideology is a kind of codification of many of the common beliefs of humankind, especially a preference for the justice of deserts compared to the justice of need and equality."[5] People think well of themselves, so they create for themselves a comfortable delusion that they are, in market terms, well valued. This reminds us that money cannot be directly proportional to happiness, since people are capable of believing all sorts of wrong things about who has what. Consider also a much-cited study of "positional" happiness, published in an article entitled "Is More Always Better?" of 1998.[6] A survey of 257 students, faculty, and staff members at the Harvard School of Public Health asked whether they would prefer to earn fifty thousand dollars a year in a world where the average salary was twenty-five thousand dollars, or one hundred thousand dollars a year where the average was two hundred thousand dollars. Half the subjects chose to be less prosperous but better off than most. This is often the only part of the study cited in the popular press, but the questions were not only about money. For example: Do you want to be the sexiest person in town, but not actually that hot, or instead would you prefer to be beautiful in a land full of absolute knockouts?

Along with income and physical attractiveness, the questionnaire asked people whether they wanted more education, or whether they wanted less but more than others; about whether they would prefer that their child was very smart among very, very smart people, or just smart but among a crowd of fools. There were also questions about vacation time, approval and disapproval from a supervisor, and how many papers you would be assigned to write. When it came to physical attractiveness and a supervisor's praise, people wanted to be ahead of other people more than they cared about the absolute amounts. When it came to vacation time, absolute quantity was more important than relative position. For money, as noted, 50 percent chose less real money if that meant they would have more than average. Praise, we may imagine, is more valuable if it sets you apart, whereas vacation time appears to have more intrinsic satisfaction. Money falls in between: it is associated with real, intrinsic goods, but it is also dependent upon relative position.

We think of money as allowing us to buy lifestyles that would be permanently conducive to happiness, but we don't appear to be all that good at guessing which lifestyles those would be. For instance, many people believe that nice weather and access to the beach can make you optimistic, relaxed, more socially open, and unburdened by coats and head colds. This, however, seems to be an illusion. Sure, we feel good when, after three days of sleet and gray, we walk out the door to find the sun shining in a warm blue sky. On the other hand, a hot town can feel grungy—sticky and peeling on one block, precious and smug on another. Warm weather is lovely, especially when you haven't had much lately, but a lot of other things can make a big difference in quality of life. In an article titled "Does Living in California Make People Happy?," research data is shown to confirm that Californians are no happier than Midwesterners.[7] If you are convinced that sun and surf are objective pleasures, remember that many island-dwelling people do not think of the beach as a place of joy and relaxation. As anyone having a bad day in Southern California knows, the symbology of bliss does not hold up against daily scrutiny. We do not guess well about what is inherently happy making.

Researcher Elizabeth Dunn and her colleagues asked college students how they felt about their soon-to-be-announced dorm assignments. They had seemingly good reasons to worry about which of the various buildings would be their home, including proximity to campus, attractiveness, reputation for noise, room size, and group amenities. But these worries turned out to be misguided. This study, published in 2003, offers a title worth reading, too: "Location, Location, Location: The Misprediction of Satisfaction in Housing Lotteries."[8] Students vastly overestimated the impact that housing assignment would have on their quality of life. According to many studies, location seems to have almost no impact on happiness. It is both intuitive and counterintuitive somehow; we knew it, and we would have never guessed it.

In 1900 about 20 percent of both men and women were at work in suits and skirts—white-collar workers. About 42 percent of men were outdoors—in farming, forestry, mining, and fishing—and the second-largest group was factory workers.[9] That same year, a third of wage-earning women were working as domestic servants—the kind of toil that no population stays in if it can escape. A hundred years later, 58 percent of men and 52 percent of women are in white-collar work. And whereas in 1900, domestic labor and farm work accounted for 47 percent of women's paid labor, in 1998 it was 3 percent. Fewer people have to survive by dangerous, backbreaking, or grossly humbling work. Consider this, too: on average, Europeans live in dwellings that have more people than rooms: a family of five in a four-room apartment. In America, it took us two hundred years to achieve the average of one room per person. We reached it just after World War II, and it took us only fifty years to double it: the typical house or apartment now has 5.3 rooms for every 2.6 people.[10] Ninety-five percent of our dwellings have central heating, whereas in the mid–twentieth century only 15 percent did. Imagine what a heated, private room for each child would have sounded like to most of history. Pure luxury.

The idea of living without these luxuries makes us unhappy. So where's the happiness crest that they should have given us? We ride it every day. We are happier than we would be if, say, two of our siblings had died from childhood diseases, and too many people lived in our house so we could never be alone, and we worked a hard job every day. It may sound self-congratulatory, but we did not make the world we live in; we inherited it from people who rejected the world they were in and drew up plans for ours. How was life before Pop-Tarts, Prozac, and padded playgrounds? They ate strudel, took opium, and played on the grass. With no time to make strudel, with opium illegal, and with the grass all paved, you will be happier if you have access to packaged cakes, modern pharmaceuticals, and foam rubber beneath the monkey bars.

This mystery holds up even on a nation-to-nation comparison. There is a correlation between a country's per capita GDP and its reported "satisfaction with one's life," but, as one research team put it, "the linkage is surprisingly weak."[11] The United States is one of the richest countries in the world, but we rank at only about the middle of advanced nations in our reported happiness.[12] Why are people surprised about this? Because they expected money to help in a way that money does not help.

Why then do comparatively well-off people keep furiously working toward wealth? Or at least fantasizing about it? Because of the abundance inference: a little food makes us a lot happier, so we think a lot of food will makes us a great deal happier still. It is hard to notice, because even though the connection between lifelong happiness and money is weak (above poverty), the short-term burst of joy is still strong. You can get some euphoria from spending money. But euphoria is spice; you can't have it for a meal. You don't want euphoria to ever start cutting into daily happiness and lifelong happiness; don't let it threaten your studies, your relationships, your job, or even your gardening. The pleasure of imagining a feast, though, is part of why we love and perpetuate the flawed abundance inference. We enjoy the idea of abundance, because the sight of it makes us feel less anxious. The way hearing a song feels good, it feels good to dwell in illusions of magnificent plenty.

Fantasies of superfluous abundance are one of the happiest things cultures have offered, and almost all cultures offer at least one. In the Hebrew Bible there is "the land of milk and honey." In medieval Europe there was the "Land of Cockaigne," where joints of meat grew on trees, macaroni fell from the sky, and pigs ran around with forks and knives in their backs, calling out, "Eat me!" This was something different from travelers' tales of wondrous places; it was a generally acknowledged fantasy, so it had no logical boundaries at all. The wild paintings of Hieronymus Bosch owe much of their imagery to tales of the Land of Cockaigne.[13] There is also the Koran's depiction of heaven overflowing with grapes and wine. (Yes, alcohol is forbidden on Earth, but handed out in heaven.) And abundance fantasy abounded in the lush European descriptions of the New World, where fish were said to leap out of the rivers. When larders were bare, heaven was sweets and meats.

Such abundance fantasy was harmless fun, not revolutionary planning. Yet it was a little subversive, especially in its rejection of religious self-denial. It was also a childlike rejection of adult reasonableness and a feminine rejection of masculine seriousness. The idea that children were more wild than adults rings a bell for us today, but we tend to depict women as the civilizing force in society, and men as big babies, led around by their immediate desires. This was not so through much of history. Self-control was often understood as a masculine virtue, and indulgence was the provenance of the female. At the feasts of carnival and the imagined festivities of the Land of Cockaigne, the women were in ascendance: there were images of wives dancing, feasting, and riding their husbands around the town square. Images of the fecund New World were often also female, and tough, fleshy girls they were, too.

Perhaps the greatest visions of abundance happiness are to be found in fairy tales. These visions of happiness are rich, bountiful, and explicitly lasting. In fact, their most consistent phrases show that they begin in a singular moment—"Once upon a time"—but that their characters experience a relief that lets them "live happily ever after." Fairy tales are only stories, but they are a mighty source of our happiness assumptions and fantasies, as well as of our conflicts and ambiguities. A look at them here will repay the extended attention.

Fairy tales were bedtime stories for children, or stories told by the old ladies while they were spinning thread, weaving, or sewing. Naturally, they are filled with the fears and hopes of mothers and their charges: worry over wicked stepmothers should the mother die, worry over the dangers and uses of beauty, worry over starvation, dreams of fortunes birthed continually from farm birds, nightmares of the hungry wolf. Because they are so often told in the moments between wakefulness and sleep, or while engaged in hypnotically repetitive work, fairy tales have the quality of dreams: they mine the subconscious, and they do not eschew the inexplicable. Because they are not included in the idea of high art, they cannot justify themselves on the basis of any of high art's fancy values. They are honed by time and necessity: if they aren't good, useful, and wanted, they get forgotten. Each individual story is proven important by virtue of its longevity.

Any fairy tale that takes a desperately poor person and gives her a home and an endless pot of food has actually earned the famous last line—and this is what a lot of fairy tales are about. Fairy tales are profoundly concerned with food. The stories originate from regular folk in medieval and early modern times, and theirs was a world that went through frequent bouts of scarcity and starvation (until steamships and railroads finally solved the problem of distribution in the mid–nineteenth century). Fairy tales feature magic tables that, on command, cover themselves in tender meats and wine; and mud pancakes that turn real. Fairy tales also harbor all sorts of ravenous creatures interested in eating little girls and boys, from

Jack's ogre atop the beanstalk, who has such an experienced nose he can distinguish the blood of an Englishman, to the old woman who lives in a candy house but prefers oven-baked children for her table. The wolf wants to eat the three little pigs, and Goldilocks wants to eat the bears' porridge. Jack's experience with the beanstalk begins when an unscrupulous person essentially seduces a naive child out of what his family needs to survive. The currency here is all food: cows, beans, bread, geese, eggs. Everyone in the story is hungry: Jack, the mother who sends him on his mission, and the ogre who wants to grind his bones for bread. The stolen golden goose will provide not real eggs, the ultimate symbol of fertility and nourishment, but rather golden eggs, extremely valuable, but barren and inedible. The golden egg is a hilarious conflation of money and food: gold, the standard of material value, that comes into being when it gets laid by a silly goose—the biological process of a domestic animal.

Everyone was hungry, children and grandmothers as well as ogres and witches, but the great hunger was imagined in the form of the wolf. In the dominant Italian and Austrian version of "Little Red Riding Hood," the wolf eats most of Grandma, then sadistically gets Little Red to drink the old lady's blood and eat her flesh, and then he eats the little girl.[14] No one comes to save them. In the 1697 Charles Perrault version of "Little Red Riding Hood," the wolf is clearly a seducer and molester as well as a vicious carnivore.[15] A wolf in Grandma's clothes offers a shadow suggestion that Grandma is hungry and will make demands. Red Riding Hood gets into her trouble in the first place because she must risk the woods to bring Grandma a basket of food.[16] The wolf may seem very male to us, but it is not so clear. In Gustave Doré's classic illustration of wide-eyed Red Riding Hood in bed with the nightgowned wolf, it looks for all the world as if the little girl is staring, aghast, not at a man-wolf in drag, but rather at a she-wolf in a nightdress. Roman slang for whore was *lupa*—literally, "she-wolf." The equation of "wolf" and "sexually voracious female" persisted in Europe into the twelfth century, and it was not until Elizabethan times that wolves became primarily symbolic of male lust. The wolf is your fear of hungers, of family hungers, the nightmare of eating, the thrust of one's own hunger, the threat of being eaten. The wolf is what gets your attention.

Wolves usually ate small mammals in the forests, but when there were no mice or rabbits, they would stalk the sheep. Little shepherds and shepherdesses might find a bloodied carcass in the morning, and if things got bad enough, the monster would come out in daylight and the youngster would see it: lips pulled back showing jagged, wet teeth; furious creased snout; shoulders muscular and huge—the whole beast weighing more than a grown man. There were numerous reports of lone wolves stalking groups of children at play, catching some littlest member, and eating him or her entirely, then and there. (Wolves can eat a huge amount at one meal and then go without for days.) When drought or cold made food especially scarce in the forests and the fields, wolves came right up to the houses. Families hid their animals and their people indoors, in defense, and that is when the wolves would start trying to get into the houses. Today we need some imagination in order to conjure up the horror the wolf represented. When you watch a movie and see a man swimming in the ocean and there is a shark fin nearby, you understand that the guy is now food, or even just an animal's plaything. It's okay, though, since humanity does not live in the water. If we did, by now only a small number of sharks would remain, and they would all be in closed-off zoological exhibits and monitored preserves. (Also, if we had built our culture in the oceans we would not have killed off the dangerous land animals, so those adventurers who made forays onto the land for pleasure, money, or science might get eaten by a wolf.) We live on land, so the die-out and containment happened to the wolves instead, but it used to be that wolves were the dry-land equivalent of sharks: at any time you could suddenly find yourself the prey of a hungry beast.

The phrase for someone living from "paycheck to paycheck," before they had paychecks, was that they were *keeping the wolf from the door*— just barely managing to keep the wolf from overpowering your defenses and eating you and your family alive. *Hungry like a wolf* describes an erotic, tumultuous hunger. A *wolf in sheep's clothing* is the very definition of betrayal; Jesus uses the term to refer to false prophets. The phrase *wolf it down* lets us imagine how it looked to watch a wolf eat. It was a terrifying spectacle. Scarcity was not only about wanting to eat; it was also about not wanting to be eaten.

Nowadays the story of the three little pigs sounds like a lesson in carpentry, but this is a story of savagery. The pigs are children. The wolf— teeth bared, drooling, and wily—is an adult. Wolves do not huff and puff and blow things down. It is a magical talent, attributable to the wolf because the monster does get in, somehow; and of course it is dramatically sexual. In the original versions, the wolf eats two pigs, and only the third brother does better. The brick house, though, is not what saves him, because the wolf is persistent enough to leap onto the roof and drop down the chimney. He almost digs the last pig out, but intelligence wins: that pig has ready a stewpot of boiling water on the fire, and the big, bad wolf himself becomes a symbolic meal.

Life is dangerous and thrilling. Parents and children compete for resources and attention, most Americans eat meat and understand its relationship to their friends at the petting zoo, and everyone wants to be so desired that they are brutally consumed. Scary, yes, but also exciting. And in fairy tales, these scary scenes are played not only to face the fear, but to indulge in the pleasure. As Djuna Barnes put it in her 1936 novel, *Nightwood*: "Children know something they can't tell: they like Red Riding-hood and the wolf in bed!" The fairy story's "happily ever after" is real: we went from being a monster's prey, to not. We once were wolf food, begging the beast to let us go, with no recourse to the law. We got rid of the wolf. Welcome to paradise.

What about the fairy stories that let a poor girl marry a prince and then live happily every after? Marriage and family seem to do people a lot of good. Also, the poor girl—Cinderella, for example—was in a bad way at the beginning of the tale: covered in cinders from tending the fire, a task both filthy and dangerous. She was poorly nourished, her friends were mice and birds, and on winter nights she was cold. At the end of the tale she is warm, fed, clean, and befriended, and likely happier, ever after. Basing our assumptions on modern research, we may surmise that ten years after the famous royal ball, she would still rate herself much happier than when she was in ashes.

"Happily ever after" made sense at the end of fairy tales because of the level of drudgery and hard work that was taken for granted at the start. Again, these were largely women's stories, and throughout history women have been occupied with the making of thread, cloth, and clothing. A little girl's first family chore was to wind the bobbins with the thread her mother had spun. If a sigh is uttered from "the distaff side" in church, it means it comes from the women's side (until recently much worship separated men and women), because the distaff was the bowlike instrument onto which women wound their thread, and they were always at it. Fairy tales are full of references to spinning, weaving, and sewing. Even the shepherdess was involved in making cloth, herding her woolen charges around the grazing fields. Women socialized at the quilting bee, where salvaged scraps of cloth were made useful again. During the endless hours that they spun masses of wool, they told each other stories, *spinning yarns*. They could make something of very low value into something of great value; in metaphor, *spinning straw into gold*. This was hard work. Painful work. One's fingers grew callused from needle and pin pricks, one's hands grew gnarled from the constant stress, one's back ached from the heavy work, one's eyes dimmed from the finer work. In common parlance, when a man came into money he would use it to save his old mother from the spinning wheel. Fairy tales show women using their cloth work to save themselves, and *happily ever after* was in part defined by no longer having to do the cloth work.

We work hard today, too; there is the stress of deadlines, the tedious commute, and the fact that many of us move paper for a living—that is, we do not personally make something lovely and needed, or provide a service that seems genuinely valued and valuable. (Such jobs are often so widely desired that for all but the elite practitioners, the pay can be low.) Most of us could not do our job outside the company for which we work, so we are also plagued by the worries of dependence. Nevertheless, the vast American middle class has it good. It is important to remember just how good it has it. What we have escaped was vicious. *Happily ever after* came into our modern world as a direct descendent of the fairy-tale ending. We have lost awareness of most of that context, but in the next chapter, "Shopping in Abundance," I will show how the largest modern public pleasure, shopping, carries the marks of fairy-tale happiness. We have seen that the abundance inference is based on a mistake in logic: the fact that the jump out of poverty gives us a true "happily ever after" does not mean that further leaps in food, stuff, and status will do anything like that for us. So we know the shopping game is flawed. We are not surprised, since the ancient sages told us that money can't buy happiness, and modern statistical studies show us that we cannot chart increasing happiness as more than a very weak correlate of increasing wealth. But there is other evidence to consider, evidence that argues quite to the contrary: our behavior. Our behavior argues that something about shopping works for us. All caveats accepted, what *is* successful about money's central game?

10

Shopping in Abundance

Money, economists like to say, is a fungible good; it is interchange-able with any other like amount of the same thing. As such it is very powerful, because it seems to stand in for all other goods. It doesn't, because money cannot be supper if there is no food at the market, and, of course, some good things are not for sale. That does not mean money cannot do anything real for us; it does, however, mean you have to trade it away to get the good out of it. Note that unlike the French, who have *les misérables*, we do not usually speak of poor people as miserables; yet people who have money but will not spend it we call *misers*. The word's first meaning was "sad," not "stingy." Those who hoard money, the fun-gible thing itself, were so well-known as miserable that they got the noun form in English: misers. There is a long history of feeling that spending is good. Note also that "spending" was the nineteenth-century euphemism for orgasm. Shopping for luxuries with money you don't have is as wrong for happiness as can be: it is taking what is not yours, it is fail-ing to control your desires, and it will make you sick with worry. And rely-ing on shopping for the greater portion of your identity—this, too, is too wrong to even bother discussing. However, for the person who has some discretionary funds, and who has an inner life, there is no reason to see shopping as a bad, soul-flattening activity. How can this, out of everything in the world, be cursèd materialism? Everything is material. Stores are just shiny. Mountain climbing—now, that is materialism. Shopping is a conceptual dance. It can make you happy.

Most shopping is done for food. Food shopping isn't usually seen as the central joy of consumerism, but let's look again. At the beginning of the twentieth century, many Americans had only recently made the leap into stable subsistence. They had left the old country, often because they were hungry. Keep in mind that it was not only about whether they had money for food: this was a world in which, often enough, there was not much food to be bought. Coming to America worked. The streets here were not golden, but it was a wonder enough that the grocer's shelves were stacked with food. This was no bony peasant behind a slate of wood with a few roots and two dead pigeons on it; it was a prosperous shopkeeper with piles and piles of produce, tiers of meats and pastries, sausages hanging from the deli ceilings, baskets overflowing with bread. New immigrants regularly wrote home listing the food available in America—apples! pears! peas! beef! duck! cakes!—and teasing their old-country relatives, "still sitting around sucking herring bones." It is reasonable to guess that such people derived lasting happiness from finally getting enough. A Czech immigrant who came to America in 1914 looked back on his life and said, "We ate like kings compared to what we had over there. Oh, it was really heaven."[1] Note the height of this joy. They were kings. They lived in heaven. For those who had come of age in scarcity, abundance was happiness.

Advertising had to try to keep this movement going beyond the first big bump, beyond the pleasure of going from having too little to having enough. The fact that advertising turned out to work so well (to even its inventors' surprise) made it possible and necessary to create yet more products. Very early in the history of advertising, the task of a food business—an industrial bakery, say—was to let women know that they no longer had to bake crackers regularly or draw a bagful from the often mouse-infested cracker barrel at the grocer's. In 1898, Uneeda Biscuits advertised their biscuits as laboratory-sterile: home was dirty; the factory was pure.[2] It seems counterintuitive to us because we think of Manchester factories belching soot, but homemakers also had imagery of local grocery stores infected with rodents, bugs, and rot. The imagery of the filthy food factory was not introduced until Upton Sinclair's muckraking novel *The Jungle* exposed foul practices in the meat industry in 1906. All

Uneeda had to do was put up a poster showing uniform, unbroken crackers in sanitary-looking, neat packages. People bought the crackers. Myriad look-alike products then appeared (many with similarly cute names, Taka Cracker and Hava Cracker among them), and advertising grew increasingly competitive.

Once the word was out that various foods were available in packages and cans, advertisers had to get you not only to choose to buy a type of product, but to buy theirs—instead of others, or also. Already have some sour pickles in the house? Why not take home some sweet ones, too? The image of heavenly cornucopia in advertising drew on the abundance inference from fairy tales. This image has been remarkably consistent in food merchandising. In the world of designer handbags and luxury cars there are minimalist stores and minimalist advertisements, but even the most exclusive food stores tend to offer a very full visual palate. What about consumerism is working for us? The image of food abundance would seem to be essential to our answer. For many people, most of them women, food shopping is nowadays not a pleasure trip, but rather a chore. Yet we cannot reasonably dismiss the positive psychological effect of visiting a place of plump, various, and plentiful food and claiming for one's basket an array of nutrients and pleasures. Satisfaction through abundance saves you from your hunger and also from the hunger of others—your spouse, parents, and children, the symbolic wolf in all of them, what they want from you and what they will not give. Abundance also appears like a gate against extrafamilial wolves, the predators that appear with poverty. Landlords and medical bills arrive at the door at times of hunger, like real wolves. Food abundance is a profound relief from all sorts of analogous tensions. Abundance is romance; romance is when there is enough of what we need to go around. In 1931 the prominent ad man James Yates told his staff: "Mrs. Jones doesn't want reality; that's a wolf at the door. She wants romance."[3] That may be so, but the wolf is also part of the romance. The Western supermarket is a fairy-tale happy ending.

Albert Brooks's *Defending Your Life* is an afterlife movie about a love affair that takes place in Judgment City, where one briefly vacations while superior beings decide whether you should go back to Earth or graduate to some higher place. At Judgment City, all the food tastes extraordinary, and no matter how much you eat, you feel good and do not gain weight. Meryl Streep's character, as the film's representative of womanhood, is elated about this. In our modern abundant world, adults can have access to a "Land of Cockaigne" amount of food every day, so heaven cannot be drawn just as a banquet anymore. Thus, for the fantasy to work, the food now has to be offered without even the consequence of feeling full! And with an even better taste! Does anyone need eggs, pasta, or pie to taste better?! Brooks had to do it because heaven is supposed to be food paradise, and nowadays, Earth already is.

Food abundance is a flawed fantasy—you don't get happy going from enough to too much—so it creates a lot of anxiety when it seems to be coming true and yet happiness is not increased. As much as the British are associated with the stiff upper lip, Americans are associated with "putting on a happy face." The song that gave us the phrase dates from 1963 and is aggressive in its repetition of the command. "Slap on a happy grin!" The song, from the musical *Bye Bye Birdie*, offers only one suggestion for how:

And if you're feeling cross and bitterish
Don't sit and whine
Think of banana split and licorice
And you'll feel fine

. . .

Just put on a happy face.

Again, we are living in the fantasy come true, and it is hard to take. How long was it before we denizens of the land of milk and honey began skimming the river down to 1 percent milk fat and offering a choice of sorbitol as a sweetener? Within a generation or two, the pleasures of the cornucopia became associated with health dangers. I'm not sure that abundance is really at fault in our corpulent bodies and clogged arteries: there have always been poor and rich fat people, and it is not certain that our present-day obesity problem is a direct result of the quantity of food now available. There was as much food for the buying in the 1950s, but people didn't eat as much. Maybe the abundance only slowly influenced us to eat more, but maybe it was something else that made us start eating more and moving around less, perhaps having to do with television, computers, and cars. In any case, the pleasure of the cornucopia became associated with health dangers, but amazingly, rather than reject the cornucopia, we devised a way to keep eating a lot by draining food of its contents. The same foods that were once limited because they were hard to get—foods rich in refined sugars and animal fats—now are available, but are to be limited as self-indulgent, erotic, and sinful. We are finally able to eat meat whenever we want. The meat beasts—cows and chickens—have been modified to loll around with great breasts, udders, and haunches, waiting for slaughter. We have got rid of most of the predators of livestock, macro and micro. Micro still presents big problems, but it used to be common for a whole population of a particular domestic animal to get sick and die, and humanity would have to do without. That doesn't happen much anymore. Yet, having succeeded in chasing off both hunger and predators, we developed an intense cultural directive to outlive our body's usual lifespan. In this pursuit we do vast, expensive tests of various foods, only to announce that red meat and milk fat are killing us from within. This new threat is of being taken over by our food—from the inside out! It is a fear of being killed by one's prey.

Still, though the supermarket in this new paradigm carries some of the killer threats of food in the past, a trip to the supermarket is always a win. Despite the rarity of its happening, we all have an image in our minds of someone winning a "millionth shopper jackpot" at the cash register of a supermarket. The unspoken feeling is that food shopping in a supermarket is always a victory, despite the real loss of capital. Food abundance is not performed at home. Its key experience is at the store, seeing it all and putting whatever you want into your basket. Because food shopping is a necessary and repetitive task, it becomes a chore. That it is widely envisioned as women's work does not help its allure. Nevertheless, we encounter the wealth of our culture when we go to the supermarket, and shopping for food recalls the primary jump from real-life scarcity to fairy-tale abundance.

Let's now turn from food shopping to the more purely symbolic ways we nourish ourselves at the market. The shopping that people mean when they say "I love shopping" tends to revolve around the purchase of clothing. Why should that be such a source of happiness? Shopping for clothes is an opportunity to show yourself and others what you are worth. New clothes suggest that you are safe and aggressive, wanted, visible to protectors, yet camouflaged from critics and predators by being one of the healthy pack. This can be absorbing work. Clothes shopping is an opportunity to sheathe yourself in something that has a market value — and, weird as it is, there are no laws about what you may choose. In the past, dressing beyond your social station has usually been illegal, or at least despised. In ancient Rome, only patricians could wear red shoes. In the medieval world, purple cloth was for royalty. Nowadays, clothes have clear symbolic value, but you get to pick the value. Choose too low, and people will wonder why you dishonor yourself; chose too high, and people will wonder why you put so much effort into your sheath. Choose studded leather, and they will know you have been threatened and are in armor. How we dress is a very subtle science. Sherlock Holmes and television crime-scene investigation agents can tell a lot from a hat fiber, but most of us read a person for the gist of him or her. Jsut lkie yuo cna raed tihs even though the letters are rearranged, you get a significant visual impression of a person without sounding out the details. Shopping is an opportunity to make choices, and to reconfirm or reinvent one's roles. Of course, cars and houses are just value sheaths that sit a little farther from the skin. Men shop for things we buy less often than once a year — big things, things that usually can be resold. Clothes shopping represents a more regular maintenance or adjustment of one's value, a grooming of one's image.

Shopping also holds our interest because it acts out some deep human dramas. Buying clothes is a celebration of freedom from the heavy tasks of keeping one's family clothed. The pleasure of shopping may seem far removed from memories of bleeding fingers at the loom and long nights spent spinning, but historically speaking, women were enslaved to these tasks only moments ago and all the way back into time immemorial. Women's clothes shopping is the enacting of an immeasurable modern triumph — the escape from endless tasks of fabric work. The first and greatest industry of the Industrial Revolution was textiles. Standard dates for the Industrial Revolution are from 1750 to 1850, and it would be another fifty years before there began a revolution of steel, chemicals, and technology. The "stuff" of industrialism, originally and foremost, was clothing. The massive proliferation of material goods grew in partnership with an explosion of advertising, carried by an explosion in media, literacy, and technology. As each grew, it expanded the others:

the stuff changed the world, and the world changed the stuff. There had long been market fairs where you could be confronted by a huge number of products at once, but even so, when we started approaching the modern consumer culture, suddenly we had to invent "world's" fairs. They started in 1851, in England, and got more and more humungous, and then essentially popped, starting in France: they spawned department stores that were permanent abundance fairs, and these ever-changing yet constant emporia dampened the need for the old fairs. Until we figured out how to make large, clear panes of glass, in the late nineteenth century, there was no window-shopping. Sidewalks came soon after, in response. Before that, when you needed to buy a dress, you went to the dressmaker; shopping was not a constant orgy of temptation. Now you strolled down the Champs-Élysées as an entertainment or went into the Bon Marché for several floors of everything imaginable, all in mirrored cases and multitiered displays. Department stores, from their beginnings to the present day, exhibit their goods in a way that highlights a dizzying, almost psychedelic profusion of goods. There are minimalist ads, and there are minimalist stores, but look what an abundance of them!

After millennia of concentration on food, it is shocking to see advertising turn clothes and things (not only food) into a dream of abundance on the order of the Land of Cockaigne. History does not show us examples of happiness as a fantasy of having many organizational gadgets and kitchen utensils. The proliferation of stuff for sale never before existed in anything like the way it does today. From the origins of advertising to the present day, many commercials are all about abundance—full of smiling, buxom women surrounded by mounds of whatever is being sold. To deal with the imagined health threat of having too much, the women in today's ads are thin as bones, but still show off their bosoms, the only part of the human body that feeds. These days breasts are even more complex symbols because many of the ones you see on television are fake: they are not big because they feed, nor are they big because the woman is herself well fed. As with our food, we found a way to be lean yet keep the image of abundance (though, with both low-fat food and low-fat breasts, the texture might be a bit off).

Shopping is an opportunity to enact one's personal power, to act out a psychologically fulfilling drama. A *People* magazine ad of 1937 (*People* was a monthly of the period) ran as follows:

There may not be as much romance in earning the dollars—someone else is usually the boss on that job—but in parting from them,
the buyer is boss. Selecting a necktie gives him a gratifying sense
of power. Buying a fur jacket is a great adventure for a woman—
she in the seat of authority, with salespeople eager to do her bidding. Yes, spending is fun. No wonder all the increase in national
income will not be spent "sensibly"—for the rarer the purchase,
the greater the adventure.[4]

Shopping is buying the wolf's pelt and becoming the wolf. Shopping is
also a matter of coping with the wolf's attentions. The more the real
wolf was no longer in the living memory of the people and the more
starvation disappeared from the normal life cycle, the more the wolf
came to refer specifically to sexual hunger and sexual danger. In the second half of the twentieth century, for the first time, lustful men were
referred to as wolves, not just in an occasional metaphor, but as a constant
nickname: "He's a wolf." Lupine hunger had to disappear entirely before
the wolf could be so simply identified with men. The "wolf whistle"
apparently dates from no earlier than 1946.[5] The Otto Preminger film
Anatomy of a Murder, which came out in 1959, had lawyer Jimmy
Stewart insist that his client's victim was worse than a ladies' man, a
womanizer, or a masher. Instead, Stewart hollers: "A wolf! A wolf!" The
word had a lot of power.

This newly purely sexual wolf was rarely entirely unwanted. In a 1953
Max Factor ad, a grown-up Little Red Riding Hood holds an open,
extended red lipstick in front of her smiling but closed mouth.[6] The tag
line reads: "To bring the wolves out." Smaller text adds that you will
wear it "at your own sweet risk" and calls the cosmetic "a rich, succulent
red that turns the most innocent look into a tantalizing invitation." Note
the explicit reference to danger; note, too, the food and blood reference
in "rich, succulent red," and the casual acceptance of desire as simultaneously innocent and inviting. The woman of the 1950s tries to get the
wolf's attention with the knowledge that he will either rip her to pieces
or become her faithful woodsman. A recent Pepsi ad featured actress
Kim Cattrall in a red riding hood and sporting CGI wolf eyes, glowing
yellow with vertical pupils. Cattrall was the forceful sexpot on the television show *Sex and the City*, so she made sense as both wolf and tempt-

ing prey. This may be an image that particularly pleases women at our moment of history, but we know that the fairy-tale wolf wearing Grandma's nightie always hinted at the sexual hunger and ferocity of women. In today's cultural discussion of fairy tales, we zero in not on the threat of being eaten, but on the gender roles of the characters. These aspects were always there, but look at the difference in what gets the central attention of the culture: The story once was primarily about death, and it is not at all about death anymore. Now it is about gender, how we struggle over personal value and agency. If buying a fur coat was once an adventure of symbolic survival, now it is an adventure in symbolic worth and control. It used to enhance her value more if he bought it for her; now she might brag that she bought it herself.

In Steven Sondheim's 1987 musical *Into the Woods,* it is Little Red Riding Hood who ends up with the wolf's pelt, and she fashions it into a cloak to wear instead of the red cape. It is a big deal, since her old identity was so completely associated with the garment. Pretty girls sometimes choose to get both tough and ugly in order to protect what they hide: the beauty in the beast. Sondheim's Little Red is not just hiding; she has put on the aggressive coat, and she remains aggressive and suspicious. The wolf had gotten her off the path to Grandma's by tempting her to go pick wildflowers in the deep woods. After that defloration, she will not let herself be tricked again. She gets fierce. It is a harsh modern solution that turns an innocent girl into a flowerless wolf. Once upon a time, a fur coat was imagined as a gift from the woman's husband; even if she did the shopping, this was his money and affection draped on her, protecting her from the cold world. Today women do all this themselves, making the money, deciding they have earned the gift, and shopping for it. Today, of course, there is the amazing added factor of empathy for the wolf. The antifur movement realizes our full mastery of the wolf. We are so in control that we feel sorry for the wolves. It is symbolic, of course, since the real thing killing animals is the sprawl of the suburbs and the malls themselves, and we have no intention of giving up those.

Here in early-twenty-first-century America, people jokingly refer to a trip to the mall as "retail therapy." Having money to spend means that either someone is taking care of you or you can take care of yourself. The money is confirming, and the performance of that confirmation is shopping. We feel that value should be rewarded, in cash when possible, and when our pockets are empty, we have a suspicion that we are not valued. Yet context is everything. If you buy a magnificent chocolate-brown leather coat with a beige cashmere lining and fur accents, is the coat more pleasurable when you are among other people with fabulous fur-trimmed coats, or when you are with people who are wearing last year's wool? Are you, on any given day, a wolf or a lamb?

All shopping allows the possibility that we will come into possession of something marvelous, graceful, and sublime, an object whose beauty reminds us of beauty at large, and the power of humanity to recognize such beauty in nature, to urge it into being in our art, and to embody it in our lives. In his 1956 prose poem "Meditations in an Emergency," Frank O'Hara wrote, "I can't even enjoy a blade of grass unless I know there's a subway handy, or a record store or some other sign that people do not totally regret life."[7] Human invention is a "sign" of hope and optimism, or at least "that people do not totally regret life." Shopping may be a dicey way to go looking for the sublime, but it is possible to find something that reminds you that other people exist and work to press the world into design. I say this in defense of the material, the manufactured, the useless, and other items for sale.

There is another reason shopping brings happiness. Wealth, democracy, and capitalism all wore down the middle-sized community in which most people in history have found meaning, status, and celebration. As we'll see in the next chapter, shopping is a way to buy back what money stole.

11

What Money Stole

There is a kind of community that is larger than your family and smaller than your country. Throughout history, family and country have been crucial realms of identity for people. They both have their problems, though: the family is small and limited, and the country is huge and vague. Complementing these powerful extremes was a medium-sized social world: town, parish, and trade (tinkers' bars, blacksmiths' guilds, glassmakers' holidays). The events of this middle-level world were as mandatory as the obligations of family and country.

At the origins of modern times, and the origins of the United States, young democracy had begun to break down all these obligations: increasingly, it was not illegal to miss church on Sunday, and increasingly, people could work outside guildlike associations and follow their own rules. The average town got big enough that people did not know each other and could not effectively monitor each other's behavior. It was all a dream of freedom. The real meaning of the "pursuit of happiness" was a bold claim that we ought to be allowed to peel off from the crowd and do our own thing. Enlightenment ideas about life suggested that individuals had value, and dignity, and had the natural right to try to be happy. The "pursuit of happiness" we find in the Declaration of Independence was a revival of an ancient idea. In private letters, Thomas Jefferson referred to himself as an Epicurean, and admired the ancient doctrine's philosophical naturalism, its secularism, and its pursuit of the happy life. Epicurus argued that a happy life was best attained in relative seclusion, with friends, but away from general human affairs, and far from politics. Epicurus ran a coed, hedonistic philosopher's retreat called the Garden, where people discussed ideas with no aim but the discussion's own pleasure. Yet even Epicurus's antipolitical philosophy is, in itself, a political expression, since it requires a government that could tolerate it. It was that kind of politics that Jefferson supported, especially as a defender of the separation of church and state.

Jefferson had other philosophical support for his involvement in politics. He was no stranger to the Aristotelian idea that a good life is one led in public governance, and that a moral life is one spent promoting the public good. Yet the Epicurean model was much more popular in Jefferson's world. Jefferson described his government service as a duty that he could not deny, but insisted his heart is best reflected by Monticello, the civilized man's country estate. For a lot of people at this time, happiness was an idleness enlivened by nature, hobbies, and study. It was retirement, and, busy as it often was, the point of it all was virtuous pleasure. The pursuit of happiness, then, was a utopian fantasy of independence, even solitude, available mostly to the rich, though since there was so much land in the New World, it still had a republican flavor.

Jefferson got the phrase "the pursuit of happiness" from George Mason, who'd been using the locution for a while.[1] Mason believed in this goal so much that he acted on it rather consistently—which is why you don't know who he is. He was enough of a founder of this country to have schools named after him, but he wouldn't leave his country estate much. History admires the Roman emperor Diocletian for retiring to a farm to raise cabbages. He was so in control of his ego that he could walk away from the most powerful position in the world in order to live out the last phase of his life among the seasons.[2] As it was in imperial Rome, so it was in the Renaissance: Petrarch lamented that wise Cicero took part in politics. Late-Renaissance humanism was community oriented, and historians invented the term "civic humanism" to wall off Petrarch's earlier, isolated version. And so it was in 1776: private happiness was lauded as a purer happiness than that found in civic service. Mason fought for the pursuit of happiness to be codified in the constitution. He lost. Though the phrase is in some thirty-three state constitutions, it never got into the U.S. Constitution. Mason wasn't there enough to win his case.[3]

Ben Franklin wrote to a friend in 1764, "By the way, when do you intend to live?—i.e., to enjoy life. When will you retire to your villa, give yourself repose, delight in viewing the operations of nature in the vegetable creation, assist her in her works, get your ingenious friends at times about you, make them happy with your conversation, and enjoy theirs: or, if alone, amuse yourself with your books and elegant collections?"[4] When asked how he lived while in France, Franklin described something similar. In a town half a mile from Paris he took walks in the garden and dined out six nights a week (Sundays he stayed home and welcomed any visiting Americans), and otherwise he played with his grandson Ben. Grandpa Ben bragged about this repose, this style of life, as if he were Epicurus or Petrarch, laying out fruitful tranquillity as the secret to a happy life. Here is George Washington, writing in 1797: "I am once more seated under my own Vine and fig tree, and hope to spend the remainder of my days, which in the ordinary course of things (being in my sixty-sixth year) cannot be many, in peaceful retirement, making political pursuits yield to the more rational amusements of cultivating the Earth."[5] Jefferson dreamed of retirement with reverence in a letter of 1796, going so far as to hope he lost an upcoming election so that "[t]he newspapers will permit me to plant my corn, peas, etc., in hills or drills as I please."[6] He did lose, to John Adams, but he accepted the vice presidency. Then he just barely beat Adams in 1800 and won handily in 1804. He wrote the following to a friend in 1810:

> I am retired to Monticello, where, in the bosom of my family, and surrounded by my books, I enjoy a repose to which I have been long a stranger. My mornings are devoted to correspondence. From breakfast to dinner, I am in my shops, my garden, or on horseback among my farms; from dinner to dark, I give to society and recreation with my neighbors and friends; and from candle light to early bed-time, I read. My health is perfect; and my strength considerably reinforced by the activity of the course I pursue; perhaps it is as great as usually falls to the lot of near sixty-seven years of age. I talk of ploughs and harrows, of seeding and harvesting, with my neighbors, and of politics too, if they choose, with as little reserve as the rest of my fellow citizens, and feel, at length, the blessings of being free to say and do what I please, without being responsible for it to any

mortal. A part of my occupation, and by no means the least pleasing, is the direction of the studies of such young men as ask it. They
place themselves in the neighboring village, and have the use of my
library and counsel, and make a part of my society. In advising the
course of their reading, I endeavor to keep their attention fixed on
the main objects of all science, the freedom and happiness of man.[7]

Jefferson here holds himself as a capable teacher of happiness, having just
given us a rundown of his day—clearly meant as a model for others.

Only by noting the historically strange amount of freedom that we
have in our lives, and our failure, of late, to take part in voluntary associations, can we understand what it is that we do with our modern wealth,
and why we want it so much. The founders' vision of the good life
includes permission to avoid the crowd, to skip church, and to put some
distance between you and your extended family—if you want. People
had been bossed around by authority; now they had won the right to
boss their own selves around. That they were going to do a better job of
it was still quite a self-conscious idea. In this move toward freedom,
bonds to country and family were reconceived, but were still strong:
people were supposed to die for their country, and value their nuclear
family with a sentimental romanticism. Meanwhile, in both de jure and
de facto ways, modern men and women were emancipated from mandatory midsize associations. You could ditch the town, ditch the parish,
ditch the family, and ditch your father's trade. Democracy and capitalism supported freedom from tradition. Yet the Enlightenment ideal was
that you would fill some of the gap with socializing of some sort—usually productive or intellectual—as is evident in Franklin's and Jefferson's
daily agendas.

One of the main observations in Alexis de Tocqueville's *Democracy
in America* (1835) was that the young United States was surprisingly
vigorous in the creation of clubs and associations: "Americans of all
ages, all stations of life, and all types of disposition are forever forming
associations."[8] Tocqueville came to believe that new associations were
essential to a democracy, especially the ad hoc variety that sees to the
local chores that, in Europe, were done by traditional local authority.
"In democratic countries knowledge of how to combine is the mother
of all other forms of knowledge; on its progress depends that of all the

others."[9] Tocqueville was impressed in 1835, but that was about the time that this sort of behavior began to decline: chores like arranging holiday events or doing quality control on local crafts were left to local government or private business, or they were allowed to disappear. Through the middle bulk of the nineteenth century, the "joining" mood further declined. It was revived, in a different form but with equal passion, around the turn of the century. The political scientist Theda Skocpol made a list of all the mass-membership organizations in U.S. history that had ever enrolled at least 1 percent of the adult male or female population. All of them—all—were founded between 1870 and 1920.[10] These include the Red Cross, the NAACP, the Knights of Columbus, Hadassah, the Boy Scouts, the Rotary club, the PTA, the Sierra Club, the Audubon Society, the Teamsters Union, and the Campfire Girls.

In the twentieth century, each generation of American immigrants brought Old World associations, and when they got here they started new clubs that referred to where they had come from. There were Italian social clubs delineating each little town of origin, and Jews had "Ferein clubs" that brought together people from a given region of the old Ashkenazi, mostly Russian-Polish, world. My ninety-five-year-old grandmother, Mollie, remembers her Ferein club fondly: they went on Yiddish theater trips together, had parties for kids and for adults, and arranged a calendar of community events.[11] Some such immigrant associations formed burial societies and loan societies, and ran foreign-language newspapers.[12] Then there were the lodges: *The Honeymooners'* Ralph Kramden was "a Raccoon," a member of the International Order of Friendly Sons of the Raccoons. The handshake involved touching elbows (first right, then left), followed by a "wooo" noise they made as they clasped the raccoon tail on their lodge hats and wiggled them, and then issued a joint declaration: "Brothers under the pelt." Even in modern sitcoms, if you have an old enough character, there is reference to Elks clubs and the like. Frank Barone of *Everybody Loves Raymond* escaped his wife, kids, and grandchildren by going down to the lodge to sit in a steam bath with other old guys.

Though such activities seem almost entirely praised, they faded away. The next generation didn't show up. We insisted that no one should make us do anything, and felt sure that our voluntary associations would always be there to fill the gap. But they are not, and now we do not get together the way we used to, and that makes us a little alienated and a little politically neutered. Most historic periods have time set aside for a certain amount of active, regular gatherings. Because attendance was mandatory—or at least societally expected—other responsibilities could not encroach upon them. Ours are envisioned as worthwhile, but voluntary and self-oriented, and that is why it is so hard for us to fit any of this into our lives.

As twentieth-century associationalism faded, the nuclear family took on much greater importance than ever before. The nation also grew in our mind's eye, as people had more and more daily information about the country. Everything in between shrank down dramatically. Life now included less extended family, shared religious observance, shared local customs, guilds, traditional clubs, and a great range of associations that met, self-governed, and drank and ate together. In almost all cultures across time and geography, the middle level of culture—the extended family and the town—had always been the most important by at least a few crucial tests. Long before people would die for their country, that bland abstract, they would die for their town, where all their relatives were buried and all their grandchildren lived. As for the nuclear family, if one family member committed an indecency and shamed the family name, even a mother (or daughter or son or father) was expected to cut ties to the offender. Can you imagine shunning your sister in defense of the honor of your aunts, uncles, and second-cousins in the eyes of your neighbors? It is historically strange that humanity is so geographically stirred up and urbanized that people do not even know their extended family or their neighbors. Now we sit in our private homes with our three other family members and we watch news about the nation.

We worked hard to get here. The midlevel was oppressive. Think of how many eighteenth- and nineteenth-century novels are about the demands of the extended family, or the town or local church. Extended family, town, and local church had moral standards (or perhaps "moralized rules") that were standing in the way of individual progress, liberty, and love. In England, the United States, France, and Russia, the great theme was how love and creativity suffer under the burdens of respectability. It was the common subject of James, Hardy, Hugo, and Tolstoy, in *The Age of Innocence, Jude the Obscure, Les Misérables,* and *Anna Kare-*

nina. Save the occasional paean to tradition, novels became associated with calls for individual freedom. Freedom from the midlevel won out. It was in the interests of democracy and capitalism. One's own nuclear family became a mini-utopia where one expected to live most of one's affective life, with a constant eye on the nation at large.

Our liberators left us in the lurch. They did not know that without mandatory associations it would be hard to keep up any associations at all. The buzz of young democracy faded in the early nineteenth century as ad hoc task-oriented community projects ended and increasingly were not replaced. At the end of the nineteenth century, and into the early twentieth, there was a remarkable return to civic activity in stable and lasting associations, and also in local, more amateurish social clubs. From the 1960s on there has been a clear trend toward social behavior that requires no commitment to actually show up anywhere on more than one or two occasions. For most people, even political action means a one-day demonstration at most—perhaps just an e-mail message to their senator or a poster in their window. It is ludicrous that these tiny acts can tag a person as one of the more publicly political members of a friend group or a neighborhood. Kids have as many clubs as ever, based on schools and sports, but the adults do not have anything of the kind. Where did everybody go?

12

How We Buy Back
What Money Stole

The rise of wealth has allowed for the development of technology and its wide distribution in the population. Most everyone has televisions and stereos; we have air-conditioning, computers, movies, and video games. Through history, even reading was a group activity, as there were not many copies of texts and not many competent readers around. People used to leave the house because they were bored and someone outside might be gossiping or playing a guitar or throwing a ball around; because the apartment was crowded; because the house was hot and the front stoop had a nice breeze (or the lodge had a good fan); to keep an eye on the kids in the street, who were out there because they were similarly bored, hot, or crowded at home; or just to flirt, or brag, or argue with someone. Trying to be entertained by other people is exhausting, though. They tease, they drone on, they criticize, they fail to show up, they borrow money and don't return it, they tell your secrets but are furious when you tell theirs. For most people, throughout most of history, there was no real choice.

In the second half of the twentieth century, for many people, staying at home became comfortable and sufficiently distracting. Many studies have shown that human beings do better when they have a lot of friends and family that they see a bunch of times a year. Yet when every family in town has a television and a computer, and perhaps even every person in every family has a television and a computer, people get isolated. We see famous people and we know who they are, but they do not know who we are. Not only does money allow people to sequester themselves in comfort, the pursuit of money forces people to work a lot rather than take part in the kind of casual leisure that builds community. Robert D. Putnam's *Bowling Alone* uses the working-class sport as a symbol of what he calls a collapse of American community, highlighting that we do still bowl, but because the leagues are gone, we now bowl alone. Putnam sees this as an alarming threat to "educational performance, safe neighborhoods, equitable tax collection, democratic responsiveness, everyday honesty, and even our health and happiness."[1] Shopping, sports, and television are what count for public fun these days. It is not exactly ironic that, having allowed us to massively desert the old associational world, abundant consumerism is now at the center of what we manage to participate in, publicly, and manage to talk about with friends and strangers alike. Media entertainment is a big part of how money stole the middle section of association. It lured us off the front stoops and porches, away from the clubs and halls. Our great public behavior now is shopping, but sports and television shows are also important in replacing our associational calendar and public conversation.

Like historical events based in religion, politics, and township, the game of watching sports has rituals, parties, heroes, and opportunities to scream your lungs out, slap hands with your neighbors, or suddenly embrace one another in a fit of joy. But that is mostly true if you go to the game, or at least to a sports bar, or to a friend's house to see the game. That is, to get a real rich associational experience out of watching sports, you generally have to leave the house, and most people don't: they watch the game at home on television. Money creates high-stakes situations out of all sorts of pastimes that were once more about social play. Once, baseball teams were chosen by location: a shop's workers and their friends. As soon as there was money in it, teams hired ringers, and soon everybody had to be a ringer. Even when the broadcast of games on the radio became popular, not everyone had a radio, so people gathered together to listen. People used to gather at the best television set in the group, too, but these days lots of people have good televisions.

Still, as an opportunity to talk to other human beings, sports talk works wonders. The fan of a team is included in a group, and the fan's team allegiance might be of use at any time. A expatriate Canadian

might be kinder on the road to a car festooned with Blue Jays paraphernalia. Sociability at the office is supposed to be secular and apolitical —
to ensure enough peace to get work done. The office, meanwhile, is
often considered off-limits for conversation with friends, since it represents toil for many, and competition for many others. Sports provides
a safe conversation topic with strangers, acquaintances, colleagues,
friends, and family members. A fan gets to take part in a mood, an identity, created by the win-loss history and highlights of the team. Even a
losing streak can make you feel chosen. When the Red Sox beat the
eighty-six-year "curse of the Bambino" and took the pennant, they were
elated, but within days newspapers and talk shows revealed that some
Red Sox fans felt lost and bereft![2] (Apparently Sisyphus, unchained,
would be angry that his rock and his mountain have been stolen from
him, and would weep for his lost identity.) Fandom can be a prodigious
force.

Television shows are the other safe subject of choice at the office and
among friends and family. People in sales may feel a responsibility to
keep up on television shows in order to have the means to interact in
this anodyne yet intimate way. Television, then, destroys community by
giving people something else to do with their free time, but it also creates a specialized common subject matter for an extremely heterodox
culture. Sometimes our attention to television breaks into a more fully
social fandom. People used to gather around the radio, and later it
became a social event to crowd together and watch *Your Show of Shows*,
I Love Lucy, and *The Ed Sullivan Show*. In the seventies, people held
weekly *Mary Hartman, Mary Hartman* parties. Today there are groups
of people who get together to watch weekly shows like *Project Runway*,
and before that *Sex and the City*, and as a more distant antecedent, the
annual Academy Awards broadcast. With *American Idol*, average people
began to take part in determining the show's course, calling in to vote
on which contestant receives a recording contract from the producers of
the show. Viewers can vote as many times as they like: this is not about
democracy; it is about dedication. The producers do not care if *the most*
people like a particular contestant. What is of essence, what "counts," is
that the contestant's fans be dedicated enough to buy a lot of merch:
concert tickets, music, and images. Those of us who are not in deep
enough to phone in a vote, or buy the music, still talk about shows with

friends and colleagues. Television shows and sports have a lot of ways of bringing us together in the great swath of human interaction that takes place from the national level all the way to the familial.

Television sports and shows are material to talk about when we are together, but they do not usually draw us out of our private homes and engage us in public behaviors. Leisure activities, the things we do because they make us happy, are often private. Our central public pleasure is shopping. People don't say *I love owning*. They do, however, say *I love shopping*. Drive into the medium-sized cities in this country on a weekend, and you see deserted streets. The people are not all at home watching television. Go to the mall, and you find everyone. Human beings have never had such opportunities for shopping. The growth of department stores was shocking to the men and women at the time it happened. Every step followed every step, literally, when the escalator changed all the rules for shopping, but, of course, I mean steps figuratively, too: First there was the invention of a product, be it casual wear or ready-made picture frames, and then the proliferation of it in quantity and variety, fed by advertising, and by production. Then there were inventions of new products dependent on the first, then wider marketing, more production, and on and on. Even department stores weren't big enough for us. In suburbia, department stores clumped together in malls, surrounded by single-note stores, all under one roof: comfortable in summer heat or winter cold. In the new indoor malls, as well as on traditional Main Streets and on great urban Broadways, shopping has grown into something truly unprecedented. The wealth of average people; the rise of advertising from almost nothing; and the massive amount of goods produced—all this is new to our moment in history.

In the first half of the twentieth century there was a lot of wide-eyed response to all the buying and all the products. In Edith Wharton's 1913 novel *The Custom of the Country*, the heroine, Udine, is a shopper. Her husband, Ralph, sees her bargaining as derived "not ... from any effort to restrain her expenses, but only to prolong and intensify the pleasure of spending."[3] Going to the shops is a controlled and contained way to have a social experience that suits our momentary desires. There are flea markets and antique stores where haggling is still an expected part of the game, but for the most part Americans have rejected that kind of shopping. Even car dealerships, one of the last bastions of the haggle, now offer the inability to manipulate the price (by you or them) as a major attraction. The last situations in which you can always expect to bargain are those where we make the only two normal purchases more expensive than cars: expert labor and real estate. The interactions that American shoppers prefer seem to be more friendly than antagonistic or wily. Some people go shopping looking forward to anonymous snippets of conversation; some go to demonstrate and share information about electronics, collectables, or fashion; some go for a comparatively personal, neighborly chat. A mob scene is not just about the advertised attraction: people also show up to be in the crowd. Some of us would never go to the mall on the day after Thanksgiving, but of course, a lot of us would. We may claim, with tautological certainty, that people go to places where there are crowds.

It also seems worth noting that neurological and psychological researchers have found a connection between the kind of new experiences that shopping provides and short-term dopamine surges in the brain. Researcher Michael Bardo, for instance, found that if you force lab rats to have a novel experience by putting them in a strange new cage, they will experience stress. But if you just make new cages available, most rats will choose novelty, and as they explore the new cage, they will experience surges of dopamine.[4] Shopping allows the exploration of new spaces and new product displays and seems to produce the same surges of dopamine. Neuroscientist David Lewis is a researcher for a London advertising consultant firm, Neuroco, and studies people's brain scans as they shop. (The shoppers wear a device.)[5] Companies want to know what makes people buy or not buy, and if they know what part of the brain lights up when people make purchasing decisions, they might be able to backtrack and figure out how to light up that part of the brain. For our interest here, it is most important just to find that (judging from these machines we have built here at the start of the twenty-first century) shopping experiences light up the brain a lot. As Lewis put it, "Shopping is enormously rewarding to us."[6] It looks like dopamine helps us to stop ruminating and take action. As a shopper hunts for an item, finds it, and makes the purchase, levels of dopamine flowing between nerve cells in the brain rise appreciably. The studies also show that once the product is purchased, chemical balances return to normal fast, so if you are shopping just to get the dopamine high, you probably experience a lot of buyer's remorse. If you actually have use for the item, or just want to have a little interaction with the store staff, you are fine; it is not as if the dopamine levels fall below their preshopping levels.

It is harder to say what exactly feels good about the hunting, finding, and purchasing. That is, many people who love to shop concentrate on one "obsession"—shoes, for instance, or model trains, innovative gadgets, lawn products, or computer equipment. The happiness they feel when they see something they want may have to do with anticipation of any number of conversations—spoken and implied. The new item poses opportunities to talk to people, and show off, and demonstrate your carefully honed preferences, your identifications. Shopping can also be an empowering experience: if you have some money, you can walk into a store and make a choice, and make something significant happen. Spending money is one of the two most adult gestures in our society, right up there with making money, and when you shop you get to demonstrate to yourself, the clerk, the shoppers, your friends, and your spouse (who may find out about it only later) that you claim the right to make these purchases.

Even those people who don't regularly shop as entertainment are still involved almost constantly in the cultural conversation about what new products are out there, and what you might want to buy. If you really don't like shopping, if you are as a vegetarian among meat eaters, you still give a certain amount of mental energy to what you reject. You may find that your refusal to wear logos, your shunning of technology, or your discomfort at department stores each offer opportunities to talk to people, and show off, and demonstrate your carefully honed preferences, your identifications. Just as we say we "love" shopping but not owning, it is rare to hear someone say, "I hate owning." By contrast, someone somewhere is probably saying "I hate shopping" right now. I think the abhorance some people have for shopping is important, in part because it helps us zero in on the salient features of the activity.

Not liking to shop could be about all sorts of things. The best reason to not like shopping is that you simply prefer doing something else. Shopping can be an activity without much creative play, intellectual stimulation, or social intimacy. There are many things to do that almost guarantee at least one of those three, so some people will always see shopping as likely to have a low happiness return. But that all depends on how you shop. A lot of what family and friends do together is shop. Many people remember back-to-school shopping trips with Mom as an event where they received an unusual amount of maternal attention: wonderful or dreaded, these experiences were intimate and foundational. Girlfriends flip through racks of clothes together and help each other choose items of appropriate social meaning. Couples shop to choose things for their home, for their kids, and for friends. There are also entirely different worlds of shopping: garage-saling, online bidding, antiquing in the country, combing through old musical recordings at estate sales and thrift shops. All of it affords the opportunity to observe people, to interact, make jokes, invent choices and combinations, learn about what people are doing, and be with someone you love. There are other ways to do all this, and some of them seem like a better bet some of the time, but that does not mean shopping doesn't work for a lot of people.

A popular reason for hating clothes shopping is that you do not want to see yourself in the mature costume of the culture. If you do not want to fit in, what's the best that can happen when you are trying things on? Either things won't fit, or they will fit and you'll feel unsettled about it. Who wants to be off the rack? Who wants to beg other people for their respect by wearing clothes with fancy labels? Valuing yourself (in the marketplace sense of choosing things that reflect your idea of your value) is not necessarily a process that equates to the cash value of an item of clothing. Some people who "hate shopping" dress themselves in a way that seems to imply a low value, or a specifically outcast status, but is really a version of claiming an elite status, by rejecting the safe, repetitive materialism of the mall. If hating shopping is so much about hating the conformity of consumerism, then loving shopping must be, to some important degree, about loving the conformity of consumerism. Many people enjoy a little bit of shopping now and again, and dress in a way that correlates well with their community and their personality. Everyone in America—be they a shopper, a shopping hater, or relatively neutral—manages their public personality in terms of brand names of various cultural meanings. A man or woman who would never wear most prominent labels forgives the issue with Levi's, since that brand name has been shaped to mean a kind of brand-neutral authenticity. Especially for shoppers, though, the experience is one of very carefully weighed communities.

Look also at the way our choices clump. Let us just think about some contemporary health terms and the kinds of meanings attached to them. We call following the experts' body rules *fitness* nowadays, though that connotes mostly exercise. Being *healthy* connotes mostly what you eat. Both influence what you wear. Told that you are a fitness enthusiast, I imagine you dressed in a shirt and pants of a single color made of a lab-created cloth, tight and covering the whole body, with a double stripe down the sides. Told you are "very fit," I see you in colorful cotton clothes, close fitting, that show a lot of skin. If the first answer to "Tell me about Julie" is "She's very healthy," it is a reasonable guess that she would not look incongruous in hemp sandals, that her hairstyle is notably simple, and that she wears little paint on her face. She may drive a hybrid car. She is likely to eat granola, though these days it is often a sugary, fatty, processed food. In fact, to imply the hemp-sandal version of "healthy," I might say that she's "very granola." Things get associated with each other that have nothing at all in common (or nothing more than any other random elements of the same world), and their values get linked, too. What associations about character go along with all this? *Healthy* suggests the person is happy; *fitness* might be associated with being a "nut" or in a "craze" and, therefore, not being happy. *Granola* connotes that the person is happy, but they might not make you happy if you are not granola yourself.

It can be dull when members list the group's characteristics as if they were a natural family of associations rather than the menu of a specific restaurant, and argue for the dishes as if they were right and other food wrong. In certain clothes you can almost guess whether a person is for or against the death penalty, and when they mention to you that they are for or against the death penalty, you can say, *Yes, I'm aware of your outfit's brief.*

The vast majority of human behavior may be understood as, in part, a separation of the sacred and the profane. All systems that guide such decisions are in part imaginary. Some are kinship based, or based on religion, or nation; some are based on things your father said; some are based on what you learned in school. By now you know what good is, and much of life is clearing away the stuff you dismiss and gathering in the stuff that is good. One efficient way to make a woman angry would be to buy her a purse that is much cheaper, or visually louder, or quieter than she usually carries. Why is she mad? Because you think that her sense of the sacred could include this profane object. Want to anger a man? Walk into the parking lot with him, pick out a wrong kind of car for him, and then ask him if it's his.

As soon as the consumerist culture got going, people noticed that uniform, machine-made, replaceable things were amazing, but soulless. The original Luddites were weavers destroying stocking-making machines that were undercutting their prices, back in 1811. But since then the idea of Luddites has always been people who reject new technology as cold and uniform. It was 1922 when Sinclair Lewis wrote in *Babbitt*: "These standard advertised wares—toothpastes, socks, tires, cameras, instantaneous hot-water heaters—were his symbols and proofs of excellence; at first the signs, then the substitutes, for joy and passion and wisdom."[7] Try the great poet Rainer Maria Rilke, writing in 1925:

> Even for our grandparents a "house," a "well," a familiar tower, their very clothes, their coat: were infinitely more, infinitely more intimate; almost everything a vessel in which they found the human and added to the store of the human. Now, from America, empty indifferent things are pouring across, sham things, dummy life.... Live things, things lived and conscient of us, are running out and can no longer be replaced. We are perhaps the last still to have known such things.[8]

Accept one more critique of the modern materialist culture. Of the postwar suburbs, singer-songwriter Malvina Reynolds wrote in 1961:

> Little boxes on the hill side, little boxes made of ticky tacky.
> Little boxes, little boxes, little boxes all the same.
> There's a green one and a pink one and a blue one and a yellow one,
> And they're all made out of ticky tacky, and they all look just the
>   same.

And the people in the houses all went to the university
Where they were put in boxes, little boxes, all the same.
And there's doctors and there's lawyers, and business executives
And they're all made out of ticky tacky and they all look just the
     same.

And they all play on the golf course and drink their martini dry
And they all have pretty children and the children go to school
And the children go to summer camp and then to the university
Where they all get put in boxes and they all come out the same.

It doesn't sound bad at all, actually. There's university, and camp, and dry martinis! And everybody gets a house! What makes it so creepy is the potential for isolation and how being in planned rows of houses together, but not enmeshed in a real social community, leads to secretiveness, outward conformity, and soul-sapping competition. The point I want to make is that these critiques are all true, but they are not the whole story. Rilke knows his fascination with transcience might have come in any age, and Reynolds might have agreed that a safe, quiet home is a good place to think and write songs.

There are ways that people get together nowadays. Locally and regularly at book clubs, for instance. Some sweat together. From football and softball to Gabrielle Roth's ecstatic dance and citywide puzzle hunts. Lots of people have national groups they belong to, which meet once a year or so. Historians get together at the annual American Historical Association Convention and some of the same ones may meet up again at the History of Science Convention. The National Puzzlers' League has an annual convention, and members might also meet up at Will Shortz's yearly crossword puzzle tournament. Once a year is hard to do, but consider what is gained in a life where one does such things some of the time. Local or distant, all are hard to maintain. We are very busy with our nuclear families, our careers, and our hunger for entertaining information. Which is fine, but you need some of this midlevel social stuff. Shopping is favored in part because it is malleable: you can do it for any amount of time, whenever and almost wherever you are. And it isn't completely unsocial.

Shopping is a chance to clothe and create oneself, and also a chance to talk to strangers, to demonstrate one's control over the immediate environment, and to bring home an object of beauty, utility, or interest. It is an opportunity to be with family and friends. It is an opportunity to be in a crowd or in exclusive solitude. We get to show the cashier that we have money and that we are big enough to use it. We get to show our loyalty to certain communities of style.

Above the poverty line, money is not the answer to happiness. Yet even in the world of the sufficient, people are wrong to say money cannot buy happiness. On a broad level, personal happiness can be enhanced by things going from good to better: living in a beautiful, safe place; having enough to share; occupying a social position where you do not have to be bullied too often. There are a lot of caveats to the idea that wealth ensures these things, and wealth brings its own problems—familial dependency, worry if one's personal worth matches one's worth as a person. Still, there are obvious happiness advantages to having some money. That's clearly true when we are talking about better food and medicine, but even with the already well off, more money can buffer you from many distinct pains. The difference between a phenomenal wheelchair and one that is just good enough is not trivial.

It is an old truism that happiness will not be explained in the treasury, but we regularly deny that with our actions. Money can buy some happiness. Consumerism, with its toys and its media and its work ethic, has stolen the midsize associations that long sustained human life, sending us all to our living rooms to eat dinner in front of the TV. But now we go to work and chat about television, and we meet each other at the shops or the mall. We communicate with each other in the symbolic associational meanings of our ever shifting wardrobes and possessions.

This chapter has concentrated on what wealth and abundance do for happiness. We used to be hounded by hunger. Indeed, eaten by the very hound of hunger: the wolf. We used to exist in three realms of society: the family, the local world, and the country. We have cut out a lot of that hunger, especially fear of being eaten on dry land, and we have cut out the demands of the local world. Happy changes, but they left us with a lot to make up for. Consumerism has become the culture's central opportunity for public performance; for being someone; and for eating and feeding, rather than being eaten.

Bodies

You are not in your body. You are your body. Your brain seems to do most of the thinking, but there is a sense in which your body is one big system, with the various parts making choices all the time; growing, changing, fighting germs, resting. Most such decisions are unconscious; many are made on the cellular level. How you feel is the aggregate of this whole system. Nobody knows how it works.

But we do know that we can influence it. After a heavy meal we feel slow, after a run in the park we feel lively, a warm bath calms us, an orgasm gives us a glow of well-being. Such dramatic happiness effects have led people throughout history to conclude that happiness is respondent to diet, exercise, sex, and spa treatments.

As for the particulars, we have less certainty. There is only one consistent message in these health theories and their historical record of effectiveness: embracing a regimen can make you happier for a while. Such regimens are almost all sold as lifelong conversions, but for most of us, that is not how it really works. Which regimen you should choose is the question on everyone's lips, but there is not much to go on. The actual choice of what you do about your diet, exercise, sex, and treatments is relatively arbitrary. Look at the track record of an assumption—for example, the idea that lifting weights can give you muscles. You may have had some brief weight-lifting experiences and found yourself unable to confirm this hypothesis, but casual evidence is so consistent that there is good reason to believe the correlation anyway. No such historical correlation exists between happiness and any given regime of diet, exercise, sex, and spa treatments. Again, what we can find consistent evidence for is that when people devote themselves to a regimen, they report happiness for a while. This happiness seems too strong a correlate to be attributable merely to a placebo effect; rather, the body responds to sharp changes in its habits.

The problem, of course, is that at any given historical moment, prevailing views about the right changes to make in how we care for our bodies can be bullying and shaming. There is a lot of widespread guilt about diet, exercise, sex, and treatments. How insistent all the advice is! Have sex a lot, don't have any sex; eat heartily, eat with restraint; hold your body in stillness, move it and have it moved; bind your ample body tightly, wear soft clothes but have no fat. Not only are the various instructions different; they are directly at odds with each other. Yet, despite our historical track record on these matters, we are today generally confident that we know best—that science and other experts have now got it right, after a long history of nonsense.

One thing that should tip us off that happiness research about the body may have some flaws is that the results are so often rejected by real people living real lives. The disjuncture between our beliefs and our actions is usually explained by saying that scientists have discovered what would make us happy, and that we are too lazy and self-indulgent to do what we are told is best. I don't believe it. We are not lazy and self-indulgent: most of us wake up every morning and go to school or work and/or feed the kids and the pets. Some of us even return phone calls. We are smart, industrious people who make happiness choices based on a lot of information, and we often choose against scientific advice. Science tells us that we would be happy if we took a brisk walk and then ate a meal of baked trout, cauliflower, spinach, and whole wheat bread, but instead we sit on the couch with our kids and eat Kentucky Fried Chicken. It is wondrous that we devote so much of our collective resources to empirical sciences of diet and exercise and then don't obey their findings. There is a big difference between the value of longevity in our rhetoric and the value we give it when we make a thousand little decisions over the course of our years.

You might decide you want to take a walk every day, but that might leave you only once a week for meditating. If you enjoy staying up late and waking up early, and you are also a fan of getting enough sleep, you are going to have to make some choices. It should not be, as Jerry Seinfeld once put it, a feud between Morning Guy and Night Guy, with your best interests represented by Morning Guy and all the power held by Night Guy: "The only thing Morning Guy can do is try to oversleep often enough so that Day Guy looses his job and Night Guy has no money to go out anymore." Most of us manage to keep Night Guy sufficiently under control so we can keep our jobs, but perhaps not much more. To wake up fifteen minutes early makes a real difference in your morning; you feel relaxed and expansive, but it means going to bed fifteen minutes earlier, and that means missing a whole half-hour show, because if you start watching, it is difficult to stop. Rather than seeing yourself as split, and having to listen to yourself lecture yourself all the sleepy day long, you might as well cop to the fact that, de facto, this is what you choose, these are your priorities. To my mind, this makes it easier to deal with our priorities: either change them or accept them.

In this section I pay some attention to happiness advice regarding health, throughout the ages, with two main goals. The first is to tease out some of the secret messages hidden in such advice. How does happiness work, that it could be associated with such diametrically opposed messages as "Live in purity and abstinence" and "Satisfy your bodily desires"? Notice that whatever is the content of our conversation about the body, we certainly seem to like talking about the body. The universe and our emotive internal lives are both vast and vague and very hard to alter. Between outer galaxies and inner dreams there exists a fleshy thing that you can really get your hands on, observe, fathom, and manipulate. That what we do with the body may be less of a science and more of a particular cultural dance should not stop us from admiring the aesthetic features of the dance, nor argue against the joy of dancing. The second major goal is to encourage happiness through historical perspective. Our experience with food, in particular, is too tense, too laden with meaning; the subject holds more than its share of cultural and personal anxiety, and this contributes to anorexia, obesity, and all the "worried well" in between. We can loosen up our experience of health instructions by remarking upon how they work.

We spend millions of dollars on health studies, and a huge amount of our time, money, and mental effort on eating right and getting exercise. Yet what really kills us is cars: motor-vehicle crashes are the leading cause of death for people aged six to thirty-three. The leading cause. After that age, crashes have to compete with heart disease and cancer, but our cars still kill a lot of us.[1] We crash our metal boxes into each other and into walls. Here we are, concentrating our money and minds on tests of whether eating a given fruit might be slightly correlated with the avoidance of some disease of old age and then trying hard to get our children and ourselves to eat the thing, carefully washing off the pesticides we read about in some other test, and meanwhile we ignore the death strips that are our roads and highways. If we rejected cars, we would have to walk, and our exercise problem would be over. So would our fuel problem and much of our pollution problem. But notice that this is completely off the radar as an option for the country. As a culture, we talk not at all about remaking the society so light rail and the Internet and the placement of schools, parks, and stores might make it very rare to have to use a car. I'm not trying to convince you of anything to do with transportation. I'm trying to say that the way we arrange for happiness—all this business with broccoli and treadmills—is not the only way we could be doing things. It is a species of madness called custom. People are going to look at us from the future with much head scratching.

What is the historical wisdom on the subject? Philosophy is often thought of as being devoted to extremes: some stodgy wise men telling you to deny the body's desires, and some iconoclastic sages telling you to enjoy the sensual pleasures. In reality, history's graceful-life philosophers tend to favor moderation. Aristotle told us that we should strive for the golden mean, to enjoy most of what the world offers, but not too often. He said the best thing one could do with one's time was to study philosophy, but he was also a man who took part in life—as a soldier, a husband, a father, a governor, a drinker, and a friend. There were ancient Greek cults of self-denial, and Aristotle dismissed them almost as firmly as he dismissed the glutton and the drunk.

Moderation also found brilliant expression in the East. The Buddha called his philosophy "the Middle Way," because, as he saw it, people seeking happiness tended to either join the monks and torture their bodies with denial, or join the worldly and go for sensual pleasures. The

Buddha said don't indulge your body in feasts, but likewise don't starve yourself. Featherbeds might be a distracting indulgence, but likewise, there is no need to go outside and sleep in the snow. Many mystics—including the Buddha—had found harsh self-denial to be very effective in reaching altered mental states, but when he came to express his own philosophy, the Buddha said that this sort of thing was a wrong turn. Your treatment of the body was about training it to an easy roughness so you could concentrate on other things. The Buddha's moderation was different from Aristotle's: for his enlightenment project, he suggested you avoid sex altogether; and he felt that alcohol and fancy food were also more of a distraction than they were worth.

Epicurus is remembered by many as the great philosopher of sensual pleasure, so associated with enjoying food that *Epicurean* today means "a lover of food." Its not an unreasonable association. As Epicurus puts it: "Pleasure is the beginning and the goal of a happy life. The beginning and root of every good is the pleasure of the stomach. Even wisdom and culture must be referred to this."[2] Fun as this is, the comment is rare among his writings. It was the medieval Christians who chose to paint Epicurus as a devotee of sumptuous feasts. They derided him because Epicureanism was one of Christianity's few important rivals in the ancient and early medieval worlds. Epicurus mentioned the pleasures of sex, drink, and food in ways that let you know he appreciated them, but he qualified their role in happiness: "Of all the things that wisdom provides to help one live one's entire life in happiness, the greatest by far is the possession of friendship. Before you eat or drink anything, consider carefully who you eat or drink with rather than what you eat or drink; not eating with a friend is the life of a lion or a wolf."[3] More conducive to happiness than fine dining or sex was thinking about what these things mean to us: "It is not continued drinking and revels, or the enjoyment of female society, or feasts of fish and other such things, as costly table supplies, that make life pleasant, but sober contemplation, which examines into the reasons for all choice and avoidance, and which puts to flight the vain opinions from which the greater part of the confusion arises which troubles the soul."[4]

One of the greatest thinkers of the Islamic golden age, the doctor and philosopher al-Razi, suggested that to promote well-being one should follow a balanced diet based on what had become "the Epicurean school": intelligent, secular pleasure. In Asia, the notion of yin and yang was very much about food and balances. Yin has a lot of sodium, yang has potassium, and you are supposed to eat a balanced meal through variation but mostly through eating things that are not too yin or too yang. The way to do that is to eat unrefined whole grain, grown as nearby as can be managed. Avoid extremely yang food, such as meat, poultry, salt, and ginseng. Extremely yin things should be avoided too: sugar, euphoric drugs, and high-proof alcohol.

Many of these calls for balance and moderation were later reimagined as extreme convictions. Unlike philosophers, most of us enjoy having something new to believe in, and find cycles of excess and self-denial easier than constant moderation. Clearly, there are pleasures at the extremes. Happiness appears in magazines as rich food and bone-thin bodies. What I want to show is that we *do* this. We create this tension to pump up these experiences, to make either eating a cookie or not eating one into an event of importance. If we could see our happiness injunctions for the body from a little distance, I think we could do better. To do that, we need to take a look at the shape of historical advice on happiness and the body, its common links and anxieties. The details here are intended to demonstrate that plausible arguments can be made for all sorts of things, and to highlight the way that such arguments travel and grow. This is about the most important thing you can learn, because science still tells a lot of "just-so stories." That is fine for science; that is how it progresses. But when it is telling us what to do, we want to have trained ourselves to recognize the difference between narrative and fact.

13

Eating

In the medieval West, on feast days, people gorged themselves in animalistic sensualism. In periods of fast they rejected food, sex, and physical effort as antithetical to holiness. Some monks and mystics renounced the pleasures of the body year-round, but for most folk, the common culture held that pleasures were good. The great Christian movements against the flesh date not from the Catholic medieval period, but from the period just after the Protestant Reformation. That is when we find sects of Christianity so hostile to the body that they refused meat and sexual intercourse. Some of them made no exception for procreative intercourse: they were trying to avoid producing more bodies, and they happily died out.

The origins of American vegetarianism can be found in William Cowherd's New Church in Manchester, England, and his issue was primarily antisex. The Reverend William Metcalfe brought New Church vegetarianism to the United States when he immigrated to Philadelphia in 1817. Metcalfe practiced homeopathic medicine and helped start the American health reform movement. As we saw, George Washington died in 1799, and that marked the beginning of a real rejection of established medicine and its bleedings and purges. Opiates were one alternative; letting nature take its course was another. Metcalfe and other reformers can be said to have exploited the vacuum in medical authority, claiming that they could make you healthy and happy through purity. They rejected opiates and advised helping nature take its course through prohibitions: no meat, no smoke, no liquor. It was Metcalfe who converted Sylvester Graham, and over the next decades, Graham forever marked the American concept of healthy living. You can't get very far in a movement that says only to reject things. Graham was first and foremost antisex, but he came up with something to say yes to, something he said would do us positive good.

Sylvester Graham championed homegrown, whole-grain, home-baked bread. If you couldn't grow your own grain, find grain farmed nearby, as local as possible; and for the magic to really work, the dough should be kneaded by the palms, fingers, and fists of the woman who gave birth to you. Again, what gives pause is not that one man would come up with such a thesis, but rather that America accepted it as a viable approach to happiness: Mom's tough-grain bread instead of sex. Graham's doctrine was powerful in part because it came at a time when there was widespread distrust of medicine and in part because it seemed to fit with the lifestyle of the times. Early American immigrants came from places where breakfast was greasy and meaty. A lot of them were farmers in the old country and they worked hard, got through cold nights, and ate lard in the morning so they could do it all again. When they came to the young United States they were astounded by the abundance of food and for breakfast set sideboards crammed with various and plainly cooked meats. As life in the United States grew more urbanized, these heavy breakfasts were called into question.

Sylvester Graham wrote extended descriptions of his mother's bread and about the taste of the grain harvested on his family's farm, and he developed a more general theory, taking in all bread, baked by all mothers. "There was," he wrote, "a natural sweetness and richness in it which made it always desirable; and which we cannot now vividly recollect without feeling a strong desire to partake again of such bread as our mothers used to make for us in the days of our childhood."[1] It is a pretty story, but there is something terribly wrong with it. Sylvester was only two years old when his father died at age seventy-four, and he was only six when the state declared his mother mad and took Sylvester out of her custody. A boy whose father dies when he is two wins the Oedipal struggle before it happens. It might seem to the baby that his desire for all his mother's attention may have killed Dad. Then, to have her taken away when the boy was six—the adult Graham was going to have some issues. This was the man who brought America's attention to the healthful properties of Mother's sweet, coarse bread and advocated total sexual abstinence. He was also against meat, alcohol, coffee, tea, opiates, and tobacco.

During the terrible and terrifying cholera epidemics of the 1830s, Graham promised protection against the disease if you were sexually abstinent and followed his grainy diet. He would go around the country lecturing about the evils of the orgasm and the blessings of bread. Sometimes these talks were so racy that he preferred to deliver them to single-sex audiences. Sometimes that wasn't enough. In the middle of a cholera scare in Boston, Graham advertised a lecture for women on protective health and was later mobbed—not for hucksterism, but for lecturing a female audience about masturbation. Graham said masturbation was bad, of course; all sexuality was, according to him. But he said altogether too much about it for some people. As we will continue to see, talking about not having sex can be just as titillating as talking about having sex; it is all the same nouns and verbs, and such lectures were certainly delivered with a kind of breathless passion. Graham cautioned that we suffered illness because we wasted ourselves in sex, and because we ate bread that was made from debauched flour.

Graham claimed that the store-bought white flour everyone used was from exhausted soil, "artificially stimulated" with animal manure. It may seem odd to worry over such a natural version of artificial stimulation, but weakened food is a very old concern. Ancient Romans were already worried about the health consequences of farming depleted soil. Even the world that made the shift from hunting-gathering to farming, ten thousand years ago, must have been impressed by this special thing being asked of farmland, and must have worried if lesser fruits would thus be grown there. In a natural orchard, much fruit falls and disappears, as do the droppings of every animal that ambles through. Now consider our strange behavior: picking all the fruit, cleaning the place up, shooing the animals, asking the land to do the whole thing again next year, and in the end taking our bodies out of the food chain. If all we do is eat, and in death we rot in parks where no fruit grows, can our farm soil be rich enough to feed us? What if the fruit is plump, but falsely so?

Clearly we did not have to wait for twenty-first-century farm practices
to start worrying. It is a valid question, and also a powerful metaphor: eating and eating, and not getting fed. Graham's disciples from around the
country created the *Graham Journal of Health and Longevity* and wrote
articles to fill it. They set up "Graham boarding houses" in Boston and
New York where young people could eat the diet, exercise, and talk about
not having sex—a subject almost as compelling as talking about having
sex, and nearly identical. Graham meanwhile became reclusive, suddenly started eating meat and drinking alcohol, and died at fifty-seven. You
would think that this would have discredited the movement, but it didn't.
His ideas about grain and internal cleanliness have never gone away.

The identification of happiness with Mom's home-baked bread may
seem like a natural idea to us, but that is only because we are used to it.
Mothers used to cook everything, and surely many of them were indif-
ferent bakers. As the century went on, packaged and canned foods
replaced much that mothers used to make at home, from scratch. One
of the last holdouts on the American dining table was bread. By the second half of the century, moms started to get that premade, too. The idea
of Mom's bread as a vision of happiness was at its most anxious and insistent when it started to disappear as a reality. *Mrs. Lincoln's Boston Cook
Book* of 1883 held that "[n]othing in the whole range of domestic life
more affects the health and happiness of the family than the quality of
its daily bread."[2] What an extraordinary and bizarre claim. After all, once
packaged sliced bread came in, in 1928, it took over fast, and many have
asserted that there has been nothing greater since. Today, most homes
do not think of the health and happiness of the family as dependant on
how good the bread is. Mrs. Lincoln, however, seemed not to doubt it in
the least. By the arrival of the twentieth century, the idea was so well
established that it was ready to be mocked. The following is from the
curmudgeonly lawyer Clarence Darrow's autobiographical novel *Farmington* of 1904:

> [Today] even the street signs ... advertise "pies like mother used to
> make." ... I cannot say that I looked upon my mother even as a
> cook exactly in the light of the street-car advertisements.... I am
> quite certain that it is only after long years of absence, when we
> look back upon our childhood homes, the bread and pies are

mixed with a tender sentiment that makes us imagine they were better than in fact they really were.[3]

For an antisentimentalist, there could be no better target than Mother's bread.

In the mid–nineteenth century, James Caleb Jackson took up Graham's antisex wheat cure and added water: hydropathy, a precursor to modern hydrotherapy. It was an aggressive incarnation of the water cure, and it usually including someone being hosed down, hard. In 1859 Jackson opened a health spa in upstate New York, featuring lectures against sex, water treatments, and, in the dining hall, vegetables and Graham bread. The spa tried to sell Graham bread in the gift shop, but it went stale too fast. Its appeal was supposed to have been its homemade nature, that it was a real organic perishable, yet capitalism is about shelf life. In 1863 Jackson cooked up a hard bread made out of Graham flour, broke it into chunks, and told people to soak the chunks in milk overnight and have the mush for a healthful breakfast. He called it Granula. It was the first prepared and packaged breakfast cereal, if not exactly ready to eat.

At around this time the pro-bread, antisex, anticaffeine, vegetarian, temperance movement found one of its greatest religious leaders. In 1864, Ellen G. White (1827–1915), cofounder of the Seventh-Day Adventists, brought a group of followers with her to Jackson's spa. White's church had grown out of a prophecy, by one William Miller, that the return of Jesus, or Advent, was going to occur in 1843. When 1843 came and the end of the world did not, the group adjusted their calculations and decided it was to happen on October 22, 1844. That date came and went, too, also without bugles from heaven. Most of the followers cut and ran, but a small group decided that something indeed *had* happened. White was the most important of these. She claimed to have received prophecies from God that encouraged her to reveal Saturday as the true Sabbath. She found Graham's diet reform inspiring and added it to her mission. Consider the song "The Health Reform," written by one of the Seventh-Day elders.

First goes the tobacco, most filthy of all,
Then drugs, pork and whisky, together must fall,
Then coffee and spices, and sweet-meats and tea,

And fine flour and flesh-meats and pickles must flee.
Oh, yes, I see it is so,
And the clearer it is the farther I go.[4]

It was not uncommon to see danger in strong flavors. In our society, pickles are considered so innocuous that it is wonderfully instructive to find them on the list of what "must flee." White and her husband moved the group out to Battle Creek, Michigan, in 1860. Having visited Jackson's spa, she reported receiving a divine mandate to open her own, the Western Health Reform Institute. White brought a strict attitude to American spas. The WHR Institute muddled along for ten years before its famed, awful, unforgettable director came along: John Harvey Kellogg.

Imagine you were raised in a culture that told you pickles were bad for you. Maybe you wouldn't believe it, but it is likely that, even so, if you made short work of a jar of gherkins you would feel a little guilty. But a lot of people raised in a world where pickles are said to be bad never even consider that the whole thing is nonsense. We have been spared this burden only by the happenstance of the roulette of time. It can be salubrious to meditate on such narrow escapes, but if that is not enough, a good look at Kellogg should do the trick.

Kellogg's family converted to White's Seventh-Day Adventist religion in the year he was born, and at the age of twenty-four, in 1876, he became medical superintendent at the spa. White sent him to a hydropathy school in New Jersey, and from there Kellogg went on to study medicine and surgery at New York's Bellevue Hospital and other notable institutions. In 1878, back at Battle Creek, Kellogg made up his own version of Granula, and sold it under the same name. Jackson sued Kellogg over the name and won, so Kellogg changed the name of his product to Granola. His wife, Ella Kellogg, wrote cookbooks inventing different ways to use the stuff. Around 1897 John Harvey and his brother Will Keith Kellogg founded the Sanitas Food Company—the name yet another example of the period's concern with cleanliness. They experimented with toasting grain mush and came up with cornflakes. By 1906 Kellogg had become too independent for Ellen White's Adventists, and she expelled him and his brother from the church.

John Harvey was the front man, and his purpose for the cereal was to help liberate people from their desire for meat and sex. I'll say more about his sexual obsessions in the chapter on sex, but here we may note that he took both desires as bestial. Feed a vicious wolf cereal mush for long enough and he will cease to bare his teeth or leap after a raw steak. The cereal scrubbed out your animal nature. Kellogg's ideas were powerful for decades; then they faded. John Harvey had wanted to keep the Kellogg movement a crusading happiness treatment, but as he got socially weirder (white suit, white Australian cockatoo), his brother Keith gained control of the company. Keith put sugar in the flakes and removed from the company literature all diatribes against meat and sex. Over the years, the advertising followed wider social trends. The sexual revolution of the 1960s discredited the idea of prescribing abstinence for happiness.

That was the rise and fall of a belief that whole grain could and should be used to help men and women to not have sex. I've never noticed in myself or others any reason to correlate vegetarianism with celibacy or carnivorousness with Eros. In fact, a nice steak can make you too greasy and satisfied to go looking for love. Furthermore, Kellogg could not display enough angry meat-eating wolves and friendly bread-eating wolves to convince me that meat makes you violent. But the display was convincing to Kellogg's audience, because it went along with the other things they believed. The concept of dry cereal and milk, and the idea that this should be breakfast, are leftovers from Kellogg's era, but they now serve new symbolic purposes in an almost wholly new trance of value.

Breakfast cereal remained a meal invested with all sorts of meaning. It is surprising to realize that what we shake out of a box in the morning has somehow managed to take on the name of its whole category, as if there were a new cheese you eat in a bowl as prelunch and we called the product "dairy." Most people in the world live on cereal, in a variety of forms. Cereals are grass with edible starchy grains: wheat, rice, rye, oats, corn, and sorghum (big in Africa and Asia). Such grains have sustained civilization in almost every case of the latter's appearance, but they have very little protein and limited other nutrients. Cereal is not the best breakfast a twentieth-century mom could envision. So in the first half of the twentieth century the cereal companies aimed marketing at children, and the sugar content increased steadily until, in 1956, Kellogg's Sugar Snaps came in at 56 percent sugar. That eventually woke people up, and the backlash caused the cereal companies to throw a bunch of vitamins and minerals into the flakes, puffs, and shreds, "fortifying" them.

In another strange turn, in the 1960s granola became so linked with body naturalism and free love that the word became a slang referent to a whole social category. Granola was seen as anticorporate and antiestablishment: a natural, unprocessed food that went with natural, unprocessed desires. As we saw, it was actually the first packaged premade breakfast cereal, invented by a company, stolen by another, but the whole grains stood out enough in 1960s America that the term *granola* was taken to mean naturalism in food and mores. Graham and Kellogg would have been shocked to see granola touted by sensualists. They might be even more surprised if they came back now and found that in a stupendous turnaround, here in the twenty-first century, breakfast cereal will make you sexy. What was once going to keep us clean and sexless is now marketed as offering youth and sexiness. Television commercials current in the 2000s show slim women getting slimmer by eating wheat- or cornflakes, ending the thirty-second spot in tight jeans or a bikini. Nowadays the flakes will get you naked.

There are remarkable similarities between nineteenth-century guilt over sex and today's worry over our body's sexiness. The levels of anxiety are the same. Worry over sex and worry over being thin enough have both generated a galactic amount of newsprint, of talk, of regimentation, of ever-revised advice. Consider the stern conviction that if we could only fully invoke our willpower—if we could only resist the temptation to pleasure ourselves—we'd be happier. And what contribution to sexual or eating experience do all our anxieties about it actually make? Did all the talk get people to stop having sex? What contribution to the eating experience have all our anxieties about it actually made? We are still eating cheeseburgers. We make a deafening buzz about how we mustn't do it, and then we go and do it more than anyone in history. Does that mean that, for all the talk, they were still having sex? Maybe a lot. Are we all just working ourselves up for the drama of it? The relationship we have to food is overwrought, too full of meaning. The whole subject is so heavy with cultural significance that it helps create both anorexia and obesity, while everyone in the middle frets and postures. And you thought ours was not an enchanted world! A witch puts a spell on a behavior, such that those who engage in it shrink to their bones or blow up like balloons. To break the spell, recall that there are vast worlds on the other sides of the body: the universe outside, and the mind within. Surely this slab of water and meat deserves less worry. Consider also this: When people counsel others or themselves with the credo "Food is not love," I reply to the television (I hear it only on television): "Oh yeah, well, not all desire is loneliness." The physiological effect of sitting down to a plate of something well made and delicious, savory or sweet, is a real pleasure in its own right, and some people like it more than others. Are the thin so jealous of the passions and pleasures of others that they cannot stop harping on longevity, though they can correlate it only with their refusals if they really search for it?

With the magic of grain, the enchanted world not only winds us up; it also sets us loose. Grain may be the site of important restrictive diets for happiness, but it is also such a strong symbol of happiness that if you want to bring gladness to a room, you need only present it. Sweetened, fluffy bread is our single strongest symbol of celebration. It is the only food that we regularly physically write on, and what we write on it is "Happy ..."! The truest birthday cake is the one that parents give their children, thanking Demeter and celebrating that they got to keep their baby. Cake is also the central totem of our "big day," the wedding. Bread is good; friendship is when we break bread together; survival is when you manage to earn your bread. Bread is happiness, and cake is even sweeter than that.

There have been many health-and-happiness crazes in history, but the nineteenth century's *fletcherizing* is among the most intriguing, important in its own time, and useful to us for its temporal proximity and conceptual distance. Fletcherizing was the commitment to chew food so thoroughly that "the food swallowed itself."[5] Wonderfully, the American Horace Fletcher met English prime minister Gladstone in London, sometime midcentury, and Gladstone told him that people ought to give every mouthful of food thirty-two chews—one for each tooth. What was Gladstone up to? It is hard to say, but note that he was part of the generation of Englishmen that was romantic about effort. Upon his return to the States, Fletcher looked at his fellow Americans and declared them fat, dull, and miserable. He went beyond Gladstone's thirty-two chews and advised that all food should be pulverized by chewing, and that whatever did not get swallowed in this process should be spat out. Indeed, it was said that Kellogg's concentration on whole grain and enemas was a result of first having been a devotee of fletcherism and having ended up constipated. This was also the presenting complaint of many of his fletcherizing spa guests. Fletcher recommended a diet of potatoes, cornbread, and beans, and an occasional egg. As important as the food choices, he taught, was that we think happy thoughts while we eat. This could not further narrow dinner conversation by much, since the chewing silenced everyone almost completely. One writer of the time commented that at dining parties "[These] strange creatures seldom repay attention. The best that can be expected from them is the tense and awful silence that always accompanies their excruciating tortures of mastication."[6] The "fletcherites" were a new "horror to dining out." Part of the point of fletcherizing was to fool yourself into eating much less than Americans were used to. Fletcher was a man of surprising physical strength and bright optimism, and he used demonstrations of his strength and personal magnetism to sell the chewing cure.

Fletcher promised that the chewing would calm you down and make you happy. He himself reported that before he discovered chewing he was in a malaise, having lost interest in "life and in [his] work."[7] The novelist Henry James attested that fletcherizing had caused his "serenity" to return, chatting with his friend Edith Wharton about "the divine Fletcher," and referring to himself as a fanatic. Henry's brother William, the Harvard philosopher, wrote in the *Harvard Crimson* that Fletcher's techniques worked wonders for the health and happiness for those who had tried it, and if they turned out to work for everyone, "it is impossible to overestimate their revolutionary import."[8] Such journals as *Scientific American* (as prestigious then as it is now) enthusiastically supported Fletcher's claims. Indeed, the magazine's articles on various health subjects took Fletcher's chewing instructions as a basic assumption.[9] Can you imagine how it would be if, on top of everything else there is to worry about, you also felt guilty about how many chews you managed per bite? It sounds awful.

Yet so many people were avid fans, reporting that since they began the extreme chewing, they felt much better. Many of us go in and out of phases of controlling our diet in one way or another. Think about what we mean when we are in such a phase and tell other people about it, with the capper that we "feel much better." What is it that feels better? Does it seem the same as what a fletcherizer felt? Our weight-reduction diets and their chewing both have an accompanying physical strain, in the belly and in the jaw, respectively, to remind you of them all day: both leave you feeling hungry, alive to your desires; both make you feel proud of your willpower; and both apparently give you the sense that you are getting happier. Whether they also create a noticeable biological "high" for a given practitioner is a matter that may be decided by judgment or trial. Try fads for happiness, but try them like a scientist. Pay attention.

Upton Sinclair was a "chewer," writing that his introduction to mastication was "one of the great discoveries of my life." Throughout his life, Sinclair wrote about poverty and hunger, and I do not think these things are unconnected. Sinclair's *The Jungle* was, to his mind, an incendiary book about American poverty. No one would publish it at first because it seemed too angry: Macmillan refused it because "[o]ne feels that what is at the bottom of his fierceness is not nearly so much desire to help the poor as hatred of the rich." This, by contrast, is how Sinclair saw it: "I wrote with tears and anguish, pouring into the pages all that pain which life had meant to me. Externally the story had to do with a family of stockyard workers, but internally it was the story of my own family."[10] A socialist journal helped get the book published. Then a funny thing happened. The public "missed" the poverty message and paid attention only to the disgusting way that the stockyard workers were told to adulterate sausage. For Sinclair, this was an evil of capitalism, and for emotional effect, he noted that this meat was headed "for the breakfast tables of America!" That's all anyone heard. The outcry led to the passage of the 1906 Meat Inspection Act and hastened the 1906 Pure Food and Drug Act. The book created no outcry about poverty. As Sinclair wrote: "I aimed at their hearts, and hit their stomachs."

I think we may do well to notice how their hearts and their stomachs are connected. A lot of campaigning for food purity is a translated worry about abundance. You still eat your fill, but you agonize over the food's contents. We are a pack of animals that allows some to have excess food while others starve. Those who have so much get finicky about what is good to eat; they become obsessed by it, re-creating scarcity for themselves so as to not feel guilty, confused, or dangerously envied. (Then they go out and preach the new scarcity to the poor. Guess what? The poor resent the lecture.) Here was a young nation faced with a bounty that had never before been conjured. To their own eyes, they had conquered hunger, and they had conquered the brutal animal world. They were civilized now, no longer rutting, gorging beasts. It was time to stop wolfing down their food. Such efforts to separate ourselves from the rest of the kingdom have always backfired a little, accidentally calling attention to the fact that we *are* animals. Beating the wolf was a kind of patricide, or regicide. In the first decade of the twentieth century, humanity looked up and saw that it had mastered the wolf. People had to cope with having won. They could not swallow it. What you can't swallow, you chew and chew.

History reveals how arbitrary and unstable the science of the body can be. Knowing this deeply can make it easier to dismiss its tyrannizing monomanias. Yet the culture of serial monomanias is itself quite stable. We like them. I want to show you how the mores and fads of people living a century ago were believed in the same way that we believe in the basic ideas behind our fads today, even though they were absurd. This suggests ours are absurd, too. What is important to happiness here is both the liberation from the particular obsessions of the culture, and the realization that we like invoking obsessions, we have fun with them, and they make us feel better for a while, until they make us feel worse. To see both of these points with a clarity worthy of their import, enjoy this short play of 1911, published in *Munsey's Magazine* and called " The Bone-Crackers: A Dietetic Comedy in One Act." It was written by the then well-known book and magazine writer Ellis Parker Butler.[11] As you read it, consider the references to our animality, and how food regimes were supposed to separate us from our animal nature. Small cuts have been made for brevity, and are indicated by italicized comments in square brackets.

SCENE: The dining room of a refined modern home, the home of people who read and think, and who are abreast of the times. The table is set for dinner, with fine china and sparkling glass. Father and mother Jones, along with Edgar Jones (his eldest son), Frances Jones (his daughter), and Will Jones (a younger son), are gathered at the table, joined by pretty Amelia Brown, the fiancée of Edgar Jones. The only food on the table is a huge platter piled high with bones.

[*There is some small chat, then:*]

MR. JONES: I met a man today who had just got back [from Europe], and he says they have no idea of hygienic eating whatever. They eat all kinds of things—sauces and all that. The only thing he saw anywhere that resembled a modern diet was among the poor peasants of southern France. They are eating boiled chestnuts.

MRS. JONES (taking a large bone crunching it between her teeth): But, father, the nut diet is not hygienic. That idea had been exploded before Amelia went abroad, hadn't it?

EDGAR (cracking a bone with his teeth): No, mother. We were eating raw vegetables when Amelia went away. It was the next year we ate nuts. Don't you remember? I wrote Amelia, asking her about the nut diet in Europe, and she said no one ate them as a steady diet there but peasants who are starving.

FRANCES (with a large bone well back between her molars): That was a silly fad, wasn't it, that nut fad? I think it was worse than the raw vegetables, although it did not give me such stomach pains. I nearly passed away when we had the raw vegetable fad! Do you remember the night we had raw parsnips, father?

MR. JONES (breaking a bone in two by setting his jaws on it and pulling down on the other end with both hands): Yes, indeed! I thought we were going to lose you, Frances. That was what finally convinced me that raw vegetables were not a rational diet. It awakened me, just as your mother's appendicitis case showed me that oat-hulls were not—(Angrily)—William!

[*William, boylike, has been gorging his food without chewing it properly, and a long, thick bone has stuck in his throat. The other end extends into the air, and his face is rapidly turning from crimson to black. Edgar reaches across the table and jerks the bone out of William's throat. It might be as well to have a sword-swallower play the part of William when this play is staged.*]

MR. JONES (still angry): William, how many times must I tell you to fletcherize your bones before you swallow them? You might as well be eating bread and butter, or roast beef, if you don't fletcherize your bones properly. Chew each bite of bone four hundred times, young man, or I'll get a jaw-meter and make you wear it!

WILL (pouting): Well, my stomach feels like a dog-fight now. It feels like—like—

FRANCES (quickly): You needn't tell us what it feels like, young man! It is no wonder it feels like it, when you won't fletcherize. . . .

EDGAR (to Amelia, in loverlike tones): Why, dearest, you are not eating? [. . .]

AMELIA (in confusion): I—no—I—I am not hungry tonight, Edgar.

[*Their discussion roams a bit here, then comes back to the bones.*]

MR. JONES: [. . .] Will you have a few more bones, Amelia?

AMELIA (politely): No, thank you. I—I haven't eaten all of these yet.

WILL (accusingly): Oh, she hasn't eaten a bone yet! . . . I'll bet she's a food-eater!

MRS. JONES (angrily): William! What are you saying? You will be calling Amelia a coffee drinker next, and then you'll be sent to bed without having your backbone rubbed. (To Amelia)—Don't mind him, dear; he is a rude, ill-bred boy. He actually scoffs at bone eating. And once—but no, that is too terrible to tell, even in a family party.

WILL (brazenly): Huh, I'll tell! I ain't afraid. Ma caught me in the barn eating a piece of white bread!

[*Mr. Jones half rises as if he could hardly refrain from taking Will in hand at once. Will grins mischievously. The whole family is shocked.*]

MRS. JONES: Will, what will Amelia think of us now? She will think we are savages!

WILL (pointing to Amelia): Well, she's a savage. She doesn't eat bones.

[*After a bit more of such talk, Amelia storms out, announcing that she is going to a restaurant. Edgar follows her and the parents peel off after them, supposedly to retrieve them, but the implication is that they will all be eating, and that, in truth, they cheat on their bone diet often, each in secrecy. Young Will says to his sister:*]

WILL: [. . .] What does father do every day when he goes out to lunch? Crack bones? He eats food! Bones? What does mother do after father leaves for the office? Does she lock herself into her room to eat bones? She eats real food. (Turning to audience.) You see! The whole world is insincere; only youth is honest. In her youth America must place her hope. The whole nation is cracking bones, but youth alone is sincere!

FRANCES: Yes! Yes! So you may have these bones, William. I am going out to be insincere.

[*Left alone, Will is served a steak as the curtain falls.*]

Note the "dog-fight" in the young son's stomach, the accusations of savagery, the consciousness and unconsciousness of their foolishness, and the final bit about the distance between what people do and what they say. All their various diets are antimeat, yet this is the very thing they cannot resist. There is sexual tension and parental overcontrol in the "jaw-meter" and in the image of the sword swallower. Notice also the parallel suggested between eating off-limits food like white bread and masturbating. Fletcher was famously known as the Great Mastica- tor. Today we may smile wryly at such a title, as if the phrase only sounds a little dirty to us; but it was their joke. This fictional family and the cul- ture at large had an inkling that it was talking about sex, just as we have an inkling that when we devote ourselves to any painful diet craze, we are in a way denying ourselves as a prayer against death. It is almost homeopathic: in order to stave off death, introduce small quantities of it into your life. The family assumes that it is eating in a way that is both hygienic and modern, despite the gnawing and chomping on giant bones, a classic sign of the dirty old king and the dirty old cur.

William James gave up the mad chewing in 1908, and Henry's doctor convinced him to do the same in 1909. In his book on religion of 1917, Upton Sinclair wrote:

In the days when I was experimenting with vegetarianism, I sought earnestly for evidence of a non-meat-eating race; but candor compelled me to admit that man was like the monkey and the pig and the bear—he was vegetarian when he could not help it. The advocates of the reform insist that meat as a diet causes muddy brains and dulled nerves; but you would certainly never suspect this from a study of history. What you find in history is that all men crave meat, all struggle for it, and the strongest and cleverest get it.[12]

In the same book Sinclair says that "man . . . is humiliated by his simian ancestry, and tries to deny his animal nature." He sighs that "[m]an is an evasive beast," trying to avoid the truth about himself: harmless sometimes, but "what are we to say when we see asceticism preached to the poor by fat and comfortable retainers of the rich?" Sinclair explains that happiness is something you can get, and that the way you get it is not by following the latest injunctions—which are a complex mess of symbols and evasions—but by knowing your history and living by the larger rules of humanity, from the ancients on. "We see so much that is wrong in ancient things," he wrote, "it gets to be a habit with us to reject them." The latest thing is not the truest, and the oldest is not automatically simple and wrong: "I have known hundreds of young radicals in my life; they have nearly all been gallant and honest, but they have not all been wise, and therefore not so happy as they might have been." Sinclair stopped fletcherizing. He died in 1968, at ninety. Fletcher never stopped chewing. He died of a heart attack in 1919, at age sixty-eight. Parents continue to urge their children to chew their food carefully, and we cannot know how much of that request is a matter of manners and how much is a remnant of the old scientific-sounding popular fantasy.

Along with the chewing, health reformers counseled a huge variety of advice and yet they were remarkably certain of themselves, never seeming to shy away from imposing some cockamamy scheme on the bodies of thousands upon thousands of real people. I do not think they were being entirely disingenuous. Many health reformers did not act like people trying to get rich; at worst we might say they sought fame as servants of society. They felt certain that people were unhappy, and they adopted a cure. How could they be so sure that one idea was better than another? It seems to me that their choices were so circumstantial, so arbitrary, really, as to press the question. Only one answer seems possible. In fact, all the health reformers were saying exactly the same thing: let us concentrate on our bodies to the occlusion of all else. The great world is too vast, the inner world too boundless. The body, by contrast, is the right size to be the locus for limitation, control, and flights of excess. Chew your food.

We have here looked at breakfast cereal and chewing, two health and happiness eating fads, one still with us, though transmuted, the other essentially defunct. We now turn for a brief look at what we believe we are supposed to be eating these days, especially this business about vitamins and minerals. Early in the twentieth century the food scientist Wilber Atwater did some lab experiments on nutritional needs and published his findings. He got a tremendous amount of attention as people celebrated finally knowing what we needed to eat for optimal well-being. But Atwater had erroneously set the daily allowance of protein and calories much too high. When researchers polled Americans, they found that by Atwater's standards, almost no one was getting enough. This crisis became the Great Malnutrition Scare of 1907–1921. Thousands of children were measured from top to bottom, including their cephalic indexes, as researchers looked for signs of trouble. They checked kids for rosy cheeks and large stature, things that were not at all stable characteristics among immigrants. National attention was obsessed with this for a generation. The *New Republic* asked this mind-boggling question in 1918: "How are we to account for the fact that malnutrition occurs as much among the well-to-do as among the poor?"[13] It was all due to a bad number, and no one thought to check that number for a while. Instead, health reformers moved on to figuring out what foods to eat in order to get the protein. Remember, if you reexamine your moment's basic knowledge, the likely result is that you either (a) waste your time confirming something "everyone knows"; or (b) discover something people are very unlikely to believe. That is how even wildly strange trances of value persist over time, trapping everyone in their moment in some absurd nonsense that no one will ever care about again.

What are vitamins? When scientists first started analyzing what was in food, they found proteins, carbohydrates, and water. That's all. They dismissed fruits and vegetables as unnecessary sugar water and fiber.

Vitamins weren't isolated and named until 1912. Elmer McCollum and others at Yale showed that an absence of vitamin A in rat diets led to rats with bad vision and stunted growth. Nutritionists across America suddenly equipped their labs with rats and started removing things from the poor rats' diets. It was quickly obvious that vitamins C and D were necessary or the subject would suffer scurvy and rickets, respectively, though it took a while to figure out how they worked. McCollum showed that the absence of vitamin B in one's diet led to beriberi. He called the foods that had vitamins "protective foods," and soon so did everyone else. Modern science has not actually gotten that much further. We have been unable to determine what quantity of vitamins people need for good health. That's why food packages list contents in terms of RDA, "recommended" daily allowances; scientists could not come up with enough of a consensus to merit stronger language.

It is not just vitamins that we are a little hazy about. Broccoli and carrots are protective against cancer, right? Wrong. Scientists have not been able to find any correlation. Why did we think there was one? The biggest reason is that in the 1970s we started doing studies of cancer worldwide and nationwide, and we found huge, manyfold differences by geography. Regarding breast cancer, researchers could draw a straight line directly relating the amount of fat in the diet to the rate of breast cancer in the population. And when people emigrated from low-cancer areas to high-cancer areas, they and their children soon had cancer rates that matched those of their new neighbors. Diet seemed the likely culprit, and this conclusion was widely broadcast in the popular press. Early follow-up studies asked cancer patients about their diets and seemed to confirm the guesses about fats and fiber. But once we started getting results from forward-looking, long-term studies (where you didn't know in advance who ended up with cancer), the correlation totally disappeared. Scientists offer this explanation: the cancer demanded an explanation ("Why me?"), and the patients provided one by perceiving that they had not had enough broccoli and carrots. Thirty years later, none of the assumptions have held up. Fat in the diet, studies found, made no difference for breast cancer. Fruit and vegetable fiber had a weak effect, or no effect, on colon cancer. Dr. Meir Stampfer, a professor of epidemiology and nutrition at the Harvard School of Public Health, explained: "People drew inferences that were in retrospect over-enthusiastic." In fact, "[y]ou could plot G.N.P. against cancer and get a very similar graph.... Any marker of Western civilization gives you the same relationship."[14] Researchers spun persuasive tales of how the beta-carotene in fruit might work as a cancer preventive, and the tales made everyone confident, but studies showed that the relationship did not exist. Studies showed that fruit and vegetables did not offer any protection.[15] Dr. Barnett Kramer, deputy director in the office of disease prevention at the National Institutes of Health, said: "Over time, the messages on diet and cancer have been ratcheted up until they are almost co-equal with the smoking messages. I think a lot of the public is completely unaware that the strength of the message is not matched by the strength of the evidence."[16]

A quarter of a century ago, Dr. Tim E. Byers, a professor of preventive medicine at the University of Colorado Health Sciences Center in Denver, had high hopes for the diet and cancer hypothesis. He recently told the *New York Times* he is "sadder now, but wiser," adding, "The progress has been different than I would have predicted." He now believes that although specific foods can affect health, diet does not seem to have a major role in cancer; he thinks food quantities, however, may make a difference. "I think the truth may be that particular food choices are not as important as I thought they were," Dr. Byers said.[17]

In February 2006, researchers completed the largest study ever to ask whether a low-fat diet reduces the risk of getting cancer or heart disease.[18] What they found was that the diet had no effect. The $415-million federal study followed some forty-nine thousand women for eight years. The result was that those assigned to a low-fat diet and those who ate whatever they wanted had the same rates of breast cancer, colon cancer, heart attacks, and strokes. Dr. Rowan T. Chlebowski, a medical oncologist at Harbor-U.C.L.A. Medical Center and one of the study's principal investigators, described the low-fat diet as "different than the way most people eat."[19] It allowed for no butter on bread, no cream cheese on bagels, and no oil in salad dressings. Michael Thun, director of epidemiological research for the American Cancer Society, said the study was so big that it is likely to be considered definitive. Of the several teams, the researchers who worked on the correlation between low fat and breast cancer seemed the least ready to give up the whole hypothesis. In the conclusion of their article in the *Journal of the American Medical Association,* they suggest that maybe eight years isn't enough, and promise to follow up on the women (though no longer be on their diet). Most experts, however, seem convinced. Dr. Jules Hirsch, physician in chief emeritus at Rockefeller University in New York City and a world-renowned investigator of diets and health, called the studies "revolutionary," adding that they "should put a stop to this era of thinking that we have all the information we need to change the whole national diet and make everybody healthy."[20] Consider how many times you have been told that fruit and vegetables do, in fact, protect against cancer. I would find it difficult to exaggerate how many times I have heard or read that. I have eaten a lot of fruits, vegetables, and fiber in order to be well and not get cancer. I want to note the moral pull that this unscientific idea has had on us, and the accreditation of value and honor to the slender broccoli eater—who, of course, is also imagined as happy. It is quite amazing.

Science has taken as one of its major tasks the tracking down of what kills us, and one result is that it comes off as a morbid scold to much of the country. Of course, if we are going to have any public health care, we have to know who is making themselves expensively ill. But note that some parts of the country are preoccupied with physical beauty and longevity, while other parts of the country see all this concern as neurotic, whiny, and pestering. These matters are not devoid of politics. We should not overdo the geographical or political lines here, but simply note that, to some degree, they exist. There are people in the country who hear advice from researchers in Boston, New York, and Washington, D.C., and feel bullied. That may sound backward to the fit, information-hungry East Coast or West Coast urbanite, but since science is so often wrong on this stuff, isn't it possible that the midcountry folk are right? Perhaps the motivation for these pronouncements is more about the cultural taste of a portion of moneyed America than it is about progress in knowledge. Moneyed America (whoever is paying for the studies and the news stories about them) wants control over life's chaos. But we don't get real power over the universe by hectoring fat people about how fat they are.

As far as what you should eat or how you should move around: everyone knows what feels healthy for them. You eat a meal with a gross amount of butter in it, you feel laden and greasy; you drink a lot, you feel poisoned; you miss sleep, you feel awful; you never move around much, it hurts to walk to the kitchen. For most people, the only really valid advice is the advice you would have assessed with your own eyes and ears, in any century, things like eat less, walk around more, get more sleep, and take pleasant poisons in moderation.

I am suggesting that these issues are overstuffed with symbolic content, that this dictates what we think about and how we think it, and that we might want to try shaking it off.

14

Exercise

Our recent ancestors made a hypothesis that physical chores were a main source of unhappiness. They developed machines that lift and transport while we sit and adjust the controls. It made us happy. Imagine that someone came into the room where you now read this book and commanded you to get dressed for the weather and do some arduous outdoor chore—fetch water, say—today and every day after. Now wake from this dream and ask yourself if you are happy to be back. At the end of a workday, for the most part, our clothes are still clean, and we seem to like it that way. There are studies that claim to show that those of us who do schedule in some exercise are happier; some even claim that the effect of regular exercise can be as dramatic as the effect of taking antidepressant drugs. But there are also studies that claim to show that exercise strains the body, damages the heart, and increases anxiety. There is a great deal of money to be made in exercise equipment and paraphernalia, and not many ways to make money by telling people that the exercise fad is a marketing ploy and a cultural myth. Media might like the story, but people are so deeply convinced that the average person ought to be doing more exercise that they worry that countering the idea would seem irresponsible. Certainly those in front of the camera are likely to be personally invested in the idea that thin, exercised people live longer and better lives, are happier, and are more virtuous.

Without looking into the issue with some rigor, educated Americans would have to conclude that the average person who exercises regularly is happier and better off than the average person who does not. We are bombarded with this message, but there is nothing like scientific proof for it. A remarkable number of the foundational, famous studies showing the benefits of exercise were conducted on tiny groups of people, ten or fewer in many cases; and in many cases there was nothing random about the sample. The physiologist Steve Blair's study, published in the *Journal of the American Medical Association* in 1989, included over ten thousand men and three thousand women, and it involved interviewing them, testing them on treadmills, and tracking their health for over eight years.[1] The strongest correlation the test found was this: the people who were in the least active fifth, the sedentary quintile, who moved into the second-lowest quintile, were the people who showed some statistically significant health and happiness gains. But as Blair explained, even with a large number of subjects, a good study would require that you take a large random sample of people and assign half of them to exercise for years on end while you assign the other half to no exercise for years on end. But since random people can obviously not be made to cooperate in these ways, there is no way of controlling for genetics and a million other factors. Still, the results we do get consistently suggest that going from no exercise to some has a significant impact on health, by many measures, but beyond that there is much less of an impact. A study at Harvard Medical School looked at seventy-two thousand female nurses, comparing exercise and heart disease for eight years. The results, published in 1999 in the *New England Journal of Medicine,* showed that women who walked at least three hours a week had a 30 to 40 percent lower risk of heart disease compared with less active women, but that women who exercised for more time or with more vigor did not see greater risk reduction.[2] What good is gym fitness, then?

Note that when we say someone is "fit," all we are saying is that the things that happen to a body when it exercises have happened to this body. With this definition of fit, athletes are fit, no question about it. But they do not live longer than average people, they are injured more, and they seem neither smarter nor happier. It is funny that the nerds lecture everyone that exercise can increase mental acuity and that this does not conflict, somehow, with their sense of the jocks as dumb. Anecdotal evidence from our own lives and the lives of people we know seems to support both sides of the question. Exercise can make some people feel good, and it makes some people blissfully happy. Yet, on a daily basis, many of us make ourselves happy by not exercising. Many people report that they do not get the blissful endorphins that they have heard about—or certainly not enough to justify the time, money, and effort. Historically, most of humanity exercised because they had to, just getting around, cleaning clothes, and heating the house. Almost all human inventions have been about saving us from physical labor. The whole history of human technology has been fueled by our preference to not exert ourselves: the wheel, the elevator, the electric hand mixer, the TV remote control. Of course you don't want to exercise.

Technology made this effortless living possible, but one still has to account for why we developed the technology. Through history, when people had the money, they got other people to clean their houses, cook meals, and chop wood. With a large gap in wealth between the rich and the poor, the rich can develop some ornate styles of life that require a lot of servants. But no matter how many poor foreigners you bring in to do your heavy work, the moment the culture offers any other opportunities, you are going to have a "servant shortage." Nobody wants to wash your underwear. Doing heavy labor for your own family can be tolerable or even nice, but unless they are quite well compensated, few people want to do it for strangers. When nobody will do anyone else's scrubbing, housekeeping standards usually fall. But in an industrial age, the situation fosters new technology, and with it a furious cycle of rising expectations. The machines raise the standards of each chore. For instance, there are tons of laundry to be done in every home today, and it must all be bright and perfect—so much so that you couldn't decide one day to do it all by hand. It would be too much work, and it wouldn't turn out right. Vacuum cleaners were supposed to be a labor-saving device, relieving you from the frequent, dirty, and heavy chore of beating out your rugs. But the arrival of vacuum cleaners made wall-to-wall carpeting feasible, and the better the vacuums got, the cleaner the rugs were expected to be. Your house is large and has only one fireplace and, furthermore, you have limited access to trees. You could not decide to chop wood to heat your home any more than you could beat out your wall-to-wall shag. You can't decide to walk to the office, either, if your commute is an hour and a half by train. Most vital physical efforts have outscaled us. The only labor available is purposeless. Stunning. We are in the midst of an energy crisis, yet when a healthy man or woman wants to exert physical effort, he or she has to plug in an electric treadmill to do it. The only available labor is not only purposeless; it is actually a drain on resources.

This chapter will explain how we came to believe that the thing we use as a metaphor for drudgery-going-nowhere, the treadmill, ought to be part of most people's lives, most days. Think about how strange it is that the same culture would invent escalators, elevators, StairMaster machines, and step classes. Or that we expect our coworkers to be clean when they get to work, and still clean at the end of the day; yet we also expect them to have a separate wardrobe for the gym, which they drench with sweat. What curious behavior. How did we get here?

In all history, the cultures most obsessed with images of bodily beauty are ancient Greece, twentieth-century Fascists, and contemporary commercial America. Such environments have a big effect on the people who live in them. Of these three, we may be the ones most interested in beauty for its own sake. The ancient Greeks exercised for happiness. Their ideal of a healthy body and a healthy mind was not just noting two goals; it was presented as a causative relationship. Their gymnasia—the word means "to exercise naked"—were the centers of their towns. In Athens, young men trained together, and as they grew up, they got rich together; it was a place to make connections. Greek art reached its pinnacle in Athens, and the favorite subject was the toned body of a beautiful young man. Sparta took strenuous exercise most seriously. There the women exercised naked, too, as muscular mothers for the next generation of soldiers. Muscle was beauty, sportive play was paramount, and pleasure was sweaty. Why? Sparta had a huge population of slaves, a captured people called the Helots, who did all the farming and other heavy labor, but who were always trying to cast off their overlords. So while the Spartans did not have to do the grinding chores of farm life, they did have to stay stronger than the slave population, which outnumbered them. Meanwhile, productive labor came to seem slavish, so that the only action for a free man, other than war, was sport and war games.

In medieval Europe, there was no "cult of the body" beyond praising plump limbs, rosy cheeks, clear skin, and clear eyes; in essence, the lack of disease. Sure, anyone very fat or very thin got teased about it, as did the very short and very tall. But there was no cult of muscle and form. In the West, from at least as early as 500 C.E. to at least as late as 1500, people were never surrounded by images of toned, naked bodies. Call forth from your imagination a man and a woman of the Middle Ages, perhaps a monk in tonsure and a lady of the court with a high, conical hat. Call forth, also, a man and woman of the early modern age: he may be a village merchant and she an alewife. Do you suppose any of them were concerned about how much exercise they were getting? They were not. They did have a lot in common with each other and with us: they associated happiness with love, sex, good food, alcohol, money, parties, faith, and games. Unlike us, they did not associate happiness with exercise. In truth, in the context of most of human history, our idea that a good life includes a lot of physical exercise is bizarre.

What most resembled exercise in the Middle Ages were the knights' tournaments. These were not invented until the most brutal, martial part of the Middle Ages was over. Tournaments became really important only when central governments grew stronger (government *is* a monopoly on the use of force) and knights no longer had an opportunity to use their skills in a constant array of bashing. These tournaments, like the Crusades, were a contained way to release the extra energies of the body politic. Nowadays we say that exercise gives us energy. Which is it? Does exercise siphon off excess energy or make us vigorous and energetic? Today exercise is championed as a way of gaining extra strength, but it is also used to release stress and pressure. It also keeps people too busy for rebellion. Today, men and women in their youth and in their prime are culturally corralled into large rooms where they run in place. Rageless, we strain against machines.

It might not have turned out this way. Early Americans resisted sport. This country was founded by Protestants, and from its beginnings Protestantism was against physical sport and games. As the first century of Protestantism drew to a close, Calvinists and Lutherans attacked the frolicking and carnival that was a big part of the Catholic calendar. Catholics danced and feasted on holidays and played in the fields on Sundays. Protestants accused Catholics of idle foolishness, for taking people away from work and defaming the Sabbath. England went back and forth from

Catholicism to Protestantism before it settled on its own rather Catholic version of the latter, so there were times when murderous tension surrounded these issues of sport. In England, pro-Catholic King James I issued the *Book of Sports* in 1618, and pro-Catholic Charles I revised and reissued it in 1633, both promoting the Sunday sports: "dancing, either of men or women, archery for men, leaping, vaulting, or any other such harmless recreation," including "may games" and other sports. When Protestants gained the upper hand, they shut these down. Catholics petitioned the king, saying Protestants were keeping them from the communal sport that kept them happy and well. The kings wanted them to have it, in part so that they would not instead drink and riot. When Anglicanism, the Catholic-y version of Protestantism, won out, the pure Protestants, or *Puritans*, were pushed out of England and brought their enthusiastic refusals to the New World. They arrived here hotly anti-Catholic and antisports.

How did America then come to love sports so? It was as in ancient Sparta. In the American South, work had a connotation of slavery, and to distance themselves from it, the masters exerted themselves in play. By the end of the seventeenth century, traditions in the American South included great displays of sports followed by town feasts. Religion and law developed along with these behaviors and supported them. Gentlemen hosted the matches and the revels of abundance that followed, proclaiming that this aggressive leisure would prepare young men for war. The rise of modern exercise came at the turn of the century, and it grew out of that same concern: the strength of the nation's soldiers. In the mid– to late nineteenth century, countries called up their young men, and governments were surprised at how scrawny these young draftees were. For England it was the first Boer War, of 1880–1881, and for France it was the Franco-Prussian War, of 1871. In the United States it was the Civil War. Oliver Wendell Holmes, Sr., reported he was sure "such a set of black-coated, stiff-jointed, soft-muscled, paste-complexioned youth as we can boast in our Atlantic cities never before sprang from the loins of Anglo-Saxon lineage." Ralph Waldo Emerson issued similar reports, and exhortations to exercise.[3] In England, Anglicans developed the idea of "Muscular Christianity" to save Christianity from asceticism and effeminacy. There, Muscular Christianity was immensely popular, but in America, Protestant opposition to sports was too strong for it at first. The dominant voice of the Victorian middle class was opposed to leisure and play, particularly among Evangelicals. But sports won. Once they stopped fighting it, Puritan and Evangelical traditions found sports to be a useful family diversion, full of models of hard work for the spiritual, financial, and military soldier.

Pre–Civil War baseball was mostly about New York City and Brooklyn, beginning with the Knickerbocker Base Ball Club in 1842. They drew up rules and found a permanent site at Elysian Fields in Hoboken, New Jersey. After the Civil War, baseball's popularity took off nationwide, not only as a diversion, but as a source of "fresh air and friendship—two things which are of all others most effective for promoting happiness."[4] In the early 1890s two YMCA employees, James Naismith and William Morgan, invented, respectively, basketball and volleyball as tightly contained court games for the training and expelling of teenage energy. In 1903 the YMCA established Camp Becket for children as "a center of happiness."

Despite late-nineteenth-century encouragement of sports and exercise, there were a lot of tensions around it, in particular for women. In the Victorian period women were supposed to avoid exercise. The idea was that women's bodies were so taxed by menstruation, pregnancy, parturition, and nursing that women had to rest a lot or they would go mad—and produce a weak new generation. The social theorist G. Stanley Hall wrote that a woman "at home with the racket and on the golf links" was ruined for motherhood. "She has taken up and utilized in her own life all that was meant for her descendants, and has so overdrawn her account with heredity that, like every perfectly and completely developed individual, she is also completely sterile."[5] The childless might be made to suffer guilt by this logic. Those happy with their fertility would be impervious to such threats, but a more subtle one was available: modern nervousness. Men, too, were thought to suffer from too much mental and physical exercise, a limp exhaustion sometimes diagnosed as neurasthenia. Theorists posited that a few individuals were "millionaires of nerve-force," but most people who did too much suffered.[6] George Beard, the "inventor" and prime theorist of neurasthenia, spoke in metaphors of the new science of electricity. As he explained it, the body has a circuit that can be overloaded:

[W]hen new functions are interposed in the circuit, as modern civilization is constantly requiring us to do, there comes a period ... when the amount of force is insufficient to keep all the lamps actively burning; those that are weakest go out entirely, or as more frequently happens, burn faintly and feebly—they do not expire, but give an insufficient and unstable light—this is the philosophy of modern nervousness.[7]

In the late nineteenth century, it seemed as if everyone was overwhelmed by the speed of things and needed a rest. Not only was energy being depleted by stress and work; people believed that the actual organs of our bodies, especially the heart, would experience fatigue if exercised often. In the year of the first U.S. marathon, 1897, the *Journal of the American Medical Association* sounded an alarm that it was "unquestionable" that the marathon would damage the runners' hearts. The concept seemed too obvious to be wrong: if you want something to last, you use it sparingly; you certainly don't go tiring it out for no reason. Textbooks said people over forty should not exercise, and that anyone with heart problems should rest in bed and shouldn't walk much or run at all.

But nervousness is a peacetime trouble, and World War I required an image of strong, healthy boys. The optimism of the 1920s made that decade a heyday of public sports. Since a certain toughness was now in vogue in the mundane world, Protestants began to break with the idea of Muscular Christianity. People like Sinclair Lewis now stood up for the spiritual version of the church. Surprisingly, Muscular Christianity, including the name itself, was then taken up by the Catholic Church. Some just wandered out of the pews and into the stands (Go, Notre Dame!). Also, as German immigration increased, another major source of sports in the young United States were "*Turner* societies," which Germans organized here "to promote physical education and disseminate rational ideas in order to advance health, happiness, prosperity, and the progress of mankind."[8] They were anti-Catholic, anti-Puritan, antitemperance—rationalists who wanted to have some fun. Sports are half controlled by those trying to find sober family fun, models of self-discipline, and heroes with "heart"; and half by those who want loud fun, beer, and models of defiance. These two trances of value do not occupy separate sections in the arenas. Extremists on both sides sometimes do, but generally speaking, we are all mixed up together in the stands. Indeed, almost all of us trace our personal history in both traditions. It burdens our experience of sport and exercise to give our subconscious such ambivalence to carry alone, so it makes a lot of sense to bring these complex allegiances to the surface of our minds and get a good look at them.

With the Great Depression and World War II, the United States faced huge and often heartbreaking challenges, but it also kept winning and growing, beyond anyone's wildest imagination. In this context, metaphors of the body as a fixed, limited system of energy were cited less and less, especially when it came to young people. Still everyone assumed that adults needed to take it easy on their bodies, and this idea went nicely with all the new appliances and conveniences that came into society in the 1950s. It also went well with martinis. It was only in the 1970s that people started arguing that you could make the heart muscle more powerful by using it a lot. Inanimate things wear down when you use them—so ran the new argument—but the heart is a muscle, and, at least in the short term, muscles get stronger when you use them (though they also might cramp or tear). The change started with the jogging craze. The physician Kenneth Cooper began jogging and crusading that everyone ought to be doing it. He made up the term *aerobics*, adding an "s" to a word that meant "living in oxygen, or living in air." There was nothing scientific about it. In his best-selling book, *Aerobics* (1968), Cooper offered a list of exercises, awarding each one a number of points, and said anyone who earned thirty points in a week was fit. He asserted that people could benefit from such a regimen at any age—in their forties, fifties, and even sixties—and that those people with heart problems should be doing more regular exercise, not less. These were shocking ideas to doctors of the time, but the message that youthful vigor was available to adults was compelling to the public. Another physician, George Sheehan, discovered jogging at age forty-five and became famous for it. He insisted on the idea of getting your heart to 120 beats a minute for half an hour, four times a week, and connected the benefits of this action not only to health but to happiness. The third great guru of running was Jim Fixx. Fixx published his *Complete Book of Running* in 1978 and captured the public imagination. His understanding of what exercise had done for him was largely mood oriented. Wrote Fixx, "I was calmer and less anxious. I could concentrate more easily and for longer periods, I felt more in control of my life ... I had a sense of quiet power."[9] The Centers for Disease Control and Prevention has been doing surveys on American exercising only since 1985, but the number of joggers in the early 1960s may have been around a hundred thousand, and by the late seventies it was over thirty million. Jane Fonda cranked it up another notch in 1982 with the release of her first workout tape. Fonda asserted that her efforts were not in pursuit of a fashionably thin body, but to make her "own body as vibrant as it can be."[10] Gyms sprouted everywhere, colossal industries of sports clothing and equipment sprang up, and since then almost everyone has, at one time, joined a gym and given the experience a good college try. Most of them gave up, but they didn't dismiss the endeavor, only their ability to rise to its challenge.

Meanwhile, a lot of time has gone by and no one has been able to prove that a certain level of exercise provides well-being and health. Jim Fixx, one of the three apostles of jogging, was out jogging when he fell down dead of a massive heart attack at the age of fifty-two. The New York cardiologist Henry A. Solomon became famous for arguing that the attempt to attain longevity through exercise was a fad based on marketing: there was no proof that a slower resting heartbeat, for instance, was indicative of well-being. Kenneth Cooper began to argue that his early calls for vigorous exercise were a mistake and that moderate walking a few times a week was enough. Jane Fonda's autobiography of 2005, *My Life So Far*, revealed to the world that she had been bulimic through most of her adult life. She says she worked out to extremes, and although exercise may have made Fonda feel good on any given day, the overall experience here was not well-being, was not happiness; it was not the most vibrant one's body could be. Fonda reports that she still suffers the damage she did to her body all those years. We have read similar attestations from such women as the singer Madonna and the *New York Times* science reporter Jane E. Brody. Brody has written of herself as having been caught up in what is now called "obligatory exercise." She explains that "obligatory exercisers often report some of the symptoms seen in athletes who overtrain.... They include anxiety, apathy, chronic fatigue, decreased appetite, depression, hostility, mental exhaustion, mood changes, changes in values and beliefs, diminished self-image, impaired concentration, emotional isolation, sore muscles and disturbed sleep."[11] Exercise is not the panacea we pretend it to be. Still and all, today the fit body is the dominant cultural anxiety: if you want to be happy, productive, and reproductive, you are advised to visit an actual treadmill. To be healthy, not go crazy, and to produce a strong next generation, everyone should exercise, "work out"—whatever that means. When we find that people who exercise more are healthy, by some measure, it might be that their health caused their exercise and not the other way around. And perhaps health causes other vibrant behaviors that we never think to ask about in some of the people who live long but are noticed only as noise in the nonexercising category. Even in what our studies do pick up, it seems that serious gym workouts many times a week are not appreciably good for you and can cause real wear and tear. We find benefit most in going from no exercise to some mild exercise, but those who do much more are not much healthier. In her *Ultimate Fitness*, the science writer and exercise buff Gina Kolata explained that as she sees things, there are three reasons people exercise: health, weight loss, and exhilaration.[12] For health, the biggest effect comes from sedentary to moving— say, from no walking to walking for ten minutes a few times a day, most days of the week. For weight loss, you can almost forget about it unless we are talking about a great deal of activity. Diet is everything; the body doesn't burn calories fast enough for moderate exercise to mean anything. As for exhilaration, Kolata explains, not everyone gets it, and those people who do get it have to work extremely hard. Moderate exercise isn't enough to bring it on. And yet, despite the ancient Greek love of the fit male form, no society in history has gone as far as ours in its equation of a good, happy life with a toned, athletic body.

What are we doing when we fill the town center with gyms? We are engaged in a combination of the two American traditions. The first is the pride of the slaveholding gentry, those who no longer do heavy labor and celebrate that fact in sport. You go to the gym to join the middle and upper class in its performance of a life so full of leisure that they have to gather in a large space and exert tremendous energy *making nothing*. The second is the religious identity that distinguishes virtue through self-limitation—those who see life as a series of disciplines where dedication guarantees success. It is a very Protestant idea that individual talent and specialness is not to be fawned over, but that all honor goes to efforts that any Joe or Jane could do, like keeping the house clean, attending church every Sunday, and abstaining from drink and profanity. In this culture there is value in dedication and the accrual of incremental rewards. Our cultural contention that exercise will make us happy is not a straightforward fact of science. Instead, there are hidden meanings.

These hidden meanings make us exercise, because we want to be the gentry, and we want to be virtuous and accrue incremental rewards. Yet even more often, the hidden meanings keep us from exercising. We cannot bring ourselves to do purposeless labor. Life may have been cleansed of hard physical labor, but most of us work as hard as ever if you judge by strain and exhaustion. Engaging in purposeless effort in addition to all that work does not always feel the way it might if we were really in charge of our lives and really at leisure. A stint on the treadmill is not really about "caring for yourself." It is about the performance of "caring for yourself" within a particular trance of value. If you like it, great. Most healthy people want to move around a little, if nothing else is keeping them from it. But the type of activity matters, of course. If you wouldn't mind going out back and chopping wood for a fire, but you don't feel like riding a stationary bike in a giant hall full of mirrors, maybe that is a very sane response to a pretty insane suggestion. Maybe you should give yourself more opportunities for purposeful exercise, or maybe you actually would enjoy the stationary bike if you kept in mind some of these issues, and divorced the physical act from its oppressive meanings.

I think we would all be better off if we did unproductive exercise only for pleasure. If we want to do exercise, we should walk somewhere we have to go to anyway, do a chore you usually get someone else to do, take the stairs, carry the baby, or chop some wood. Forget the gym unless you love it, or perhaps need a change of habit. These exhortations to be a certain type of body are the nonsense jabbering of history. Anyone above the lowest quintile of activity is not going to get happy as a direct result of exercising. If you are exercising and do not enjoy it, or are not exercising and spend time feeling guilty about it, I recommend that you find something to occupy yourself that you do enjoy, whether or not it gets your heartbeat up.

Most important, remember that it is reasonable to believe that exercising keeps you young and that it gives you energy, and it is also reasonable to believe that if you want your heart to stay "like new," you should use it gently. That goes for the rest of your body, too. As with a baseball mitt, you want to keep it supple, but you do not need to go out of your way to wear it out. Likewise, it is reasonable to imagine we have a fixed amount of energy in any given day and any given life, and we should be careful not to give too much of our energy to running around. We base our lives on the opposite assumption. Both are equally sensible. Perhaps I should say that I believe in science, but that, paradoxically, the person who believes most in scientific progress believes least in scientific knowledge, because the conviction that we will progress assures me that much of what we now know will someday soon be proved wrong, or be considered totally off the point. The fact that something makes perfect sense doesn't mean it is true. We can learn to hear the difference between the kind of science stories that tend to be stable for centuries and those that change every decade.

The human sciences change the fastest because they are about humans, and our ideas about humans change a lot faster than our ideas about the sky. Evolution is stable. The evolutionary "just-so stories" that explain the origins of our particular modern behaviors are not stable. Often science must use "just-so stories," to get ahead of itself and then double back and check the dependant hypotheses. But I don't have to believe them. Doesn't it essentially have to be true that you can wear yourself down by exercising, and also give yourself strength by exercising? In some ways it gives you more life and in some ways it gives you less. Some eras prefer martinis. There are traps in all singular interpretations of the body and its place in the world (or its place on your couch). There are lots of great metaphors; sometimes they are useful, and sometimes misleading. Noticing that is a fast way out of our common trance.

Even the most scientific of the happiness instructions has to pass the test of being interesting if it is going to get on the front page. Newspapers can't sell copies headlining the same commonsense information over and over; scientists and reporters both need for the advice to keep changing. But it is obvious, it seems to me, that you should eat food that agrees with you, that makes you feel good, that doesn't lead to illness. Munching grassy foods all day and then having to take a pill for the gas, going to a restaurant expecting to leave in pain—these are not rational decisions. Why do you want to eat all that spinach anyway; and the rest of you, why take on that oversated pain? Is it about the food, or what the food means? I'd say the meaning thing. In the first case it is a matter of chewing one's way to a mythic vision of lifelong happiness. In the second case it is an aggressive choice of good-day happiness over the others, a dyspeptic overstuffing in response to the parsimoniousness of life. If you are tired, you should sleep more. But, shy of eating things and portions that cause you pain or damage your ability to breathe well or take a walk, the rest of health and happiness is overdetermined. There are so many factors influencing them that the variables win against the brain; we really cannot say what specific things "caused" the outcome.

Even if you fail to follow the best advice in a whole range of ways, your successes may outweigh your failures: you may eat junk food but be free of the eroding effects of stress. We have seen studies showing the longer lives of Americans who are female, married, rich, white, exercising, not red-meat eating, educated, red-wine drinking, stress free, and optimistic. So if I'm a married African-American man, successful but not educated, very stressed but optimistic, and I rarely exercise, eat red meat, or drink red wine, how am I doing? How much does it matter if I was breastfed, or live near a highway, or have a dog? What if you are unusually beloved?

Researchers test the health effects of prayer at a distance(!) but could not begin to tell you about the life-sustaining properties of different kinds of love. Through history women have allowed doctors to dose them to help them manage the frustrations that go along with the joys of doing full-time childcare. That sounds bad, right? Yet women live longer. Is it the drugs, or the love? You need a computer—metaphorically, a microscope—to see the longevity effects of regular exercise, but women have been obviously outliving men, in plain sight, and we all assume it is because of an absence of what the men have: the men have work stress, so they die faster. Maybe it is the other way around. After all, there is no such thing as "stress." We made it up in order to talk about things. It is a metaphor. It may not be the most helpful metaphor.

As a final thought here, consider that at the end of the nineteenth century and into the twentieth there were birthrate scares all over Europe and the United States. Women in industrialized countries were having fewer babies. (The populations were still growing, but more slowly than before.) In places where the birthrate seemed to be slowing as compared with other countries — France, for example — there was real hysteria over the matter: huge government conferences, magazine articles, whole journals on the question, a fortune spent researching the matter and trying out methods to coax women to have more babies. The effort never had much of an impact on how many babies women had. Now it is out of fashion to talk about such things. It is visible as racist, since world population is, of course, problematically large, not small. But it also just seems rude for the government or any group to stick its nose into the marital bedroom like that. Instead, it is acceptable to harangue the country about weight loss and exercise. This would have seemed rude a century ago. Such scares are poignant enough to hold our fears for us and they pass in and out of style.

15

Sex

With all this instruction about food and exercise, where is the modern lecture on how and when to have sex? Sex is an integral part of what many people are interested in when they talk about diet and exercise, but in our particular moment in history there is very little expert instruction about sex itself. Three millennia of sexual duties and restraints, and now we're supposed to do as we please!

Or are we? Today's expert advice is that consenting adults should do what they like, though we will think you happier if you have some sex rather than none, and if you do not seem to put too much effort into that part of your life. What odd criteria for normalcy. We tend to see much effort in these matters as a sign of pathology or at least "problems," even if you protest you are having a great time. Note how little of this sex instruction has to do with particular acts or a particular schedule.

For that kind of detail we have placed the emphasis on sexy rather than sex, on youth rather than potency. What we look for in today's sciences of bodily happiness is youthful vitality. Sex is today mentioned as a proof of that vitality, rather than the more traditional other way around, where all vitality is an indication of sexual potency. That's a big cultural flip. I'm not saying we are not interested in sex, but the way we express it is quite unusual indeed: we talk endlessly about sexiness and what we are doing to get sexy, but we judge each other happiest when our sex lives are almost invisible and unmentioned, indicated, perhaps, by a wry smile.

Searching for happiness, men and women throughout history have been advised to address the issue through their sexuality: through fulfillment, abstinence, or monogamy, and a million further details. To address happiness through sexuality is timeless, and, in contrast, the particular ideas that frame sex at any given moment in history are likely to be peculiar to that moment. If you are one of the many people who at some point in life feel sexually abnormal, note that a century ago, a heterosexual married couple with cosmopolitan, secular values, having good sex three times a week, might well have felt shame and anxiety over it. The couple might well have driven themselves nuts with worry, thinking they were depleting themselves in this behavior. They might have consoled themselves that three times was not that much, but they might never notice the larger assumption, the idea of our bodies having a limited amount of energy. Some ideas become visible only the moment they disappear. A century ago, an average man who had not had sex in three years might have felt proud of his health and forbearance, and a woman might have praised herself for the health and happiness benefits of ten years of abstinence. Today sex is understood as a happiness requirement. There is no large category of people—widows, spinster aunts, priests, nuns, monks, and old scholars—who are expected to somewhat blithely forgo the whole thing.

Meanwhile, as I said earlier, to be considered happy, one is supposed to keep one's sex life invisible and unmentioned. People guess that you are happy when you are in a committed relationship or, at certain ages and situations, in a period of serial monogamy. Anyone whose sex life becomes visible, due to a lot of partners or a lot of talk about specific acts, is not considered happy. Imagine a zaftig gourmet who also has several sex partners. Our culture tells us that she is sad. In his own day, and for centuries after, England's King Henry VIII seemed self-indulgent and boorish, but, century after century, the assumption was that he was having a good time. Today, if you like to eat, you are filling an inner void that, presumably, thinner people do not have. Likewise, if you have a lot of intimate encounters, you are seen as quintessentially alone, closed off to other people. So it is the well-fed one who is empty! And the one with all the partners is lonely! Of course, there is some truth here, and all historical periods have caught a glimpse of sadness in extreme behaviors—but today's rendition is pretty heavily skewed toward this peculiar, inverted interpretation of things. Most historical eras have guessed that a big part of the reason that Casanova was a Casanova was that he was good at it, that he could do it. For us it's a syndrome. As strange as it is that we pathologize anyone who has lots of sex, it may be even stranger how we associate sexual happiness with not talking about sex or doing anything noticeable about it.

There are periods in history that abound with rules about sex. There have been countless lists of instruction about what people ought not to do, and there have been many forms of social pressure about what people ought to be doing. You are likely aware of some of the strict religious opinions on the subject, and of some people's extreme habits: from conjugal sex through a hole in the sheet to metal chastity belts. Let's look at an equally weird behavior, but closer historically: the mid 1800s to the mid 1900s in America. I want to show you the reasoning of the people of those times, the way they made their arguments, because though their convictions were very different from those dominant today, the weakness of the logical links and the bold certainty of the assumptions is familiar indeed. The key sexual anxiety of the second half of the nineteenth century, in America, was one we hardly entertain at all today: worry over the practice and effects of masturbation. Homosexuality was barely discussed by comparison. In fact, homosexuality was often mentioned only because it could mean men showing boys how to masturbate. Pundits and parents did not justify or excuse homosexuality. Rather, they ignored it—essentially the way masturbation is ignored in our culture today. Why despise masturbation? The line in the Bible about onanism is tiny, and rejects an instance of birth control, not solitary sex.[1] The real classical worry about masturbation was biological. When a farmer gelds a rooster, or a bull, or a dog, the animal grows less belligerent and softer of cheek—in short, more feminine. Why? Well, what was in those things? Sperm, right? The quite reasonable assumption was that sperm, sitting in the testes and reabsorbed into the body, made a man a man. Cut off the testes and the creature will grow womanly. We know now that the testicles also produce testosterone, and that this hormone, not reabsorbed sperm, triggers the development of masculine qualities. Still, we can comprehend the old logic: dump too much semen and you will have the same effect as gelding. The subject lent itself well to worry over masculine strength. Of course, an occupation so solitary and intense is a ready site for anxiety, but women's self-pleasuring went largely unmentioned.

The modern terror began in the 1750s, when Swiss physician Simon André Tissot put together his *Treatise on the Diseases Produced by Onanism* from a variety of odd diatribes gathered from across history. Tissot's explanation of why masturbation was so harmful was as follows:

> A person perspires more during coition than at any other time. This perspiration is perhaps more active and more volatile than at any other time: it is a real loss, and occurs whenever emissions of semen take place.... In coition it is reciprocal, and the one inspires what the other expires. This exchange has been verified by certain observations. In masturbation there is a loss without this reciprocal benefit.[2]

A most incredible deduction. The effects of the "loss" were the symptoms of syphilis and gonorrhea; or just weariness, weight change, and irritability. At worst, the whole human race was going to degenerate. In Tissot's schema, women were at risk, too, since they too suffer from expiring life force in sweat and breath, and need to breathe it in from a partner to equal things out. Tissot's book was translated and published in New York in 1832 and found its way to John Harvey Kellogg. As we saw earlier, Kellogg was raised in the pro-grain, anti-sex movement, but Tissot gave him what felt like proof. Kellogg took Tissot's doctrine one step further: shared sex was almost as bad as masturbation. Once again, why did the doctrine catch on? For one thing, people feel tired a lot. After sex (alone or otherwise) a lot of people fall right asleep. They guessed that sex causes general tiredness, too.

Kellogg was severely motivated against masturbation and was vicious about it, recommending "cures" that included surveillance, special clothing, and even surgery. In his words:

> A remedy for masturbation which is almost always successful in small boys is circumcision. The operation should be performed by a surgeon without administering an anesthetic, as the brief pain

attending the operation will have a salutary effect upon the mind, especially if it be connected with the idea of punishment. In females, the author has found the application of pure carbolic acid to the clitoris an excellent means of allaying the abnormal excitement.[3]

Kellogg was against all sex, even bragging that he and his wife never consummated their marriage. He counseled that parents should surprise at night even the most innocent-seeming child and check for moist genitals. Since the vice could cause so many enfeebling symptoms, any enfeebling symptom was a sign of the vice. Acne and sleeping a lot were understood as caused by masturbation. It was an unkind use of calling a correlate a causation. In his historical study of Kellogg, John Money reveals that he himself had been raised according to these principles, and that Kellogg's ideas made his young life awful. When he decided to write a book on Kellogg, Money was director of the Johns Hopkins Medical School Psychohormonal Research Unit, professor of medical psychology, and professor of pediatrics, emeritus. In the preface to his otherwise impersonal scholarly work, Money explained the origins of his work on Kellogg:

> This book was conceived … on Tuesday, October 6 1981, [when] my cousin Meredith Money, who is my own age, showed me an old "doctor's book" he had retrieved from his mother's estate. It was J. H. Kellogg's *The Ladies' Guide in Health and Disease*.… My cousin thought I would be interested in what the author had to say about vicious habits and solitary vice in the section on "The Little Girl." Indeed I was! Here in my own hands I held, and with my own eyes I read, the very words that had shaped the sex-phobic child-rearing policies of the generation that had reared the both of us.[4]

I am trying to convince you that people really believed such things. And there are things about our present-day beliefs that will seem just as outlandish to the future. Which beliefs? Most of them. It matters that you notice how culture works and not be its slave.

All this anti-sex talk wasn't really a flight from sex. Kellogg managed to use anti-sex to create an experience that was at least what we might call "displaced eroticism." At his Battle Creek spa, people were hosed down by strong men and palpated by stern and pretty women. Clients came to him because they thought themselves too thin (too fat was a less frequent complaint, since that was still mostly a sign of wealth and health), and he would feed them hearty diets and, as a treatment, strip them down and cover them in sandbags to force their bodies to accept the nutrients. Also, Kellogg was interested in enemas of all sorts and prescribed aggressive innovations in this realm. In the movie *The Road to Wellville,* based on the novel by T. Coraghessan Boyle, you can see Bridget Fonda's character on the receiving end of a nervous treatment sometimes called a *womb massage,* which was essentially a hand job. The film is about the Battle Creek Sanitarium, but her character leaves the spa to get this treatment, and the character of Kellogg, played by Anthony Hopkins, harshly condemns the treatment when he finds out. I wanted to know what the real Kellogg thought of it, so I obtained a copy of his manual, *The Art of Massage,* and turned to "Massage of the Special Regions." Here's a bit of what I found:

> With the patient lying upon her side ... the operator stands facing the back, with the fingers resting upon the buttocks, and manipulates the perineum, using the thumbs in alternation, stretching the tissues away from the median line. Only one thumb should be used at once.... The patient should also be made to execute breathing movements, in which both the abdominal and the perineal muscles are vigorously contracted during the act of expiration. Under the instructions of a physician, the manipulations may be somewhat extended and varied by introducing the forefinger into the vagina or rectum, the muscle being grasped between the forefinger and the thumb and thoroughly pressed and stretched.[5]

Happy now? The manual is full of such instruction: fingers inserted into vaginas and rectums, the patient told to pump her hips rhythmically while taking deep, corresponding breaths. Breast massage included gently stroking, caressing, and nipple pinching. Various marketed vibrators are discussed. Pelvic massage for men included massage of the prostate; for women, "one finger placed inside the vagina"; inside the rectum and coccyx for both. Lubrication is recommended. Pregnant women needed relief, too. For them, "massage of the vagina" should be done, daily, with "strong vibratory movements," during the last six or eight weeks of pregnancy.[6] Kellogg preached against sex, but you ended up naked a lot, with a relative stranger interacting with your privates. At lunch you could look around the cafeteria tables and know that everyone else was being poked, penetrated, and prodded, and they knew that you were, too. You and the other lovely society people would share information about your treatments, and about the vices you were there to conquer. Then off to a coed lecture on what your body goes through during sex. For many years people flocked to Kellogg's Battle Creek Sanitarium, often spending full summers there. John D. Rockefeller stayed there, as did Theodore Roosevelt. Kellogg wrote with glee of turning his "despairing throng" of clients into "happy and useful people."[7]

In 1972 the American Medical Association declared that masturbation is normal and in no need of medical management.[8] As we saw in the chapter on diet, the Kellogg company had already, by that time, separated its breakfast cereals from antimeat, antisex, and antimasturbation messages. Advertising needs to track the culture's sensibility closer than doctors do. We live in a period that is remarkably approving of sex in general; yet, while Kellogg raged against the body's desires and we speak of them as healthy, he talked a lot about actual genital sex, whereas we speak mostly of sexiness. In our discussions of sexiness you will hear about secondary sex characteristics—breasts, hips, body hair, soft skin—but the old antisex sermons were replete with dewy, engorged, blushing genitalia, and descriptions of the hot exhalations of the body in its rhythms. In antisex speeches, the body sighed, heaved, gave way, sweated out salts and oils, perfumed the air with its potent musk. We say we now approve of the sexual body, but look at how we hover at a distance.

Sex advice has often been presented as having a goal other than happiness, especially advice to not have sex in order to be good, or advice to have sex in order to have babies, but a great proportion of sex rules are presented as keys to happiness. The Buddha said sex was not sinful in any way, but that it got in the way of enlightenment. As we saw, he suggested that the way to nirvana, ultimate bliss, was to have no sex at all. By contrast, other Buddhist sages have disagreed and seen sex as a path to enlightenment. Tantric sexual practices use the euphoria of sex to enhance mystical experience, and vice versa. The Hebrew Bible describes sexual devotions as a means of happiness and holiness. Consider some erotic lines from the Song of Solomon. A woman is speaking to and about her lover:

> I would cause thee to drink of spiced wine of the juice of my pomegranate.
> His left hand should be under my head, and his right hand should embrace me.
> I charge you, O daughters of Jerusalem, that ye not awake my love, until he please.
> Who is this that cometh up from the wilderness, leaning upon her beloved?

Kabbalah, the Jewish mysticism that developed in the Middle Ages, describes mystical joy as sexual, including descriptions of the male and female aspects of God making love to each other, shaking the world with their bliss. Catholicism famously celebrates virgins and celibacy vows, but it is also true that the centuries are filled with descriptions of the erotic ecstasies of nuns. Nuns are celibate and isolated, but officially they are not maidens at all. They are brides of Christ—and they have recorded intense nocturnal encounters with him. Protestantism, too, has some history of seeing sex as a godly approach to happiness; Luther said that ministers should marry and reject abstinence, and he himself married the ex-nun Katharina von Bora.

Still, let's face it: the more established a religion gets, the more its best interests are served by conservatism, and that often means that religion supports containing sexuality, requiring celibacy, sex only in marriage, limitations on when sex is allowed, and mandating which sex acts are off-limits. Eighteenth-century Enlightenment philosophers were jauntily positive about sex, in part because religion seemed so against it. French Enlightenment philosophers connected being free with the body to being free in the mind, and they often used reason and philosophy to talk women out of their knickers.[9] The "utopian socialist" settlements of the nineteenth century were economic experiments, but many of them, like the Fourierists and Saint-Simonians, also included a degree of free love. There were also communities set up primarily as pro-sex movements. The best known American sex settlement was the Oneida commune, which flourished in late-nineteenth-century New York. Its founder, Alfred Noyes, asserted that everyone in his group was married to everyone else. They all lived together and had special sex rooms for arranged meetings. It was Noyes who came up with our now common phrase "free love," though he usually preferred the term "complex marriage." The community's original forty-five members grew to seventy-two by February 1850, two hundred fifty by February 1851, and was over three hundred in 1879. In that year, though, the practice of mutual marriage was ended, partly in response to the protests of the wider community, and partly because Oneida community parents found themselves protective of the way that their teen-age daughters were being initiated into the sex life of the community. What you are not expecting me to say is that one of the group's most sacred convictions was that men should have no orgasms except for procreative purposes, to preserve their personal power. This was considered a more holy way to have sex (again, just for the men—the women could orgasm) and was touted as saving couples from the strain and woe of pregnancies. Noyes's wife's four miscarriages influenced this decision. Teenage studs were generally paired with postmenopausal women, so one could practice and the other grin. Overall, the men got variety, and the women got men trained to take their time. Both got a scheduled reprieve from the most powerful detriment to any model of adult sexuality: children. I tell you this story in order to lean a bit on our idea of sexual mores as progressive or regressive. Sure, things go up and down, but we forget that they also change ladders. Nobody seems to plan utopias anymore, and it seems to me that they don't because they do not really believe that things will ever change. Yet that is the one thing I can assure you. Things will change.

Of course, most people of the late nineteenth century were much more conventional than Noyes and his marital mates, but as the new century approached, Victorian mores began to break down. One champion of the new sexual ideas was physician and researcher Havelock Ellis. From the 1890s through the first decades of the new century, he was the leading sex expert arguing for sexual pleasure—a much respected, even beloved, cultural character who weighed in on the progressive side of all the moment's hot debates. Ellis was part of the Fabian Society, along with Annie Besant, George Bernard Shaw, H. G. Wells, Sidney and Beatrice Webb, and a handful of others, a political and intellectual think tank whose members supported gradual (nonrevolutionary) social and sexual liberation. Ellis's six-volume *Studies in the Psychology of Sex* (1898–1928) was published to enormous controversy, as were his *Sexual Inversion* (1897), a positive study of homosexuality, and his *Erotic Rights of Women* (1918). Ellis was a real hero to a lot of people. It is common to find testimonials to him as a life-changing influence against guilt, frustration, and ignorance.

Marie Stopes was also a crucial figure in this effort. She is now perhaps best known for her work in making birth control available, but Stopes was once most famous for her book *Married Love* (1918), which was really about married sex. How did she get away with it? Well, she took precautions. The book's preface begins: "More than ever today are happy homes needed. It is my hope that this book may serve the State by adding to their numbers."[10] Although she does not return to this idea of justifying her sex book by calling it a service to the state, she often returns to the idea that her book's purpose is happiness. Stopes explained that although an unmarried woman might not feel sexual needs, a wife who has been made love to but not brought to orgasm would be cranky, nervous, and sleepless. A husband should be attentive to what gives his wife pleasure and orgasm, and take the time and effort to provide it at every sexual encounter.[11] Furthermore, a married couple should follow the wife's cycle of sexual desire, which Stopes said could be generally approximated as "[t]hree or four days of repeated unions, followed by about ten days without any unions at all." Continued Stopes, "I have been interested to discover that the people known to me who have accidentally fixed upon this arrangement of their lives are *happy*."[12]

She argued that although men certainly could lose their vitality by having too many orgasms, they did not need to hold them back, but merely to refrain from spilling seed other than according to the prescribed schedule. If men or women get too little or too much sex, they get sick: neuralgia, "nerves," and growths. By contrast, she explained, "mutually happy marriage relation" leads only to positive healing and vitalizing power.[13] She wasn't forthcoming with details, but still, Stopes could be erotic in her writing; we read of the couple's "fusion of joy and rapture," the "half-swooning sense of flux which overtakes the spirit," and the "mutual penetration into the realms of supreme joy." With her era, she assumed we each had a limited sexual economy, and had to negotiate it: "This truth should never be lost sight of in a marriage; where between the times of natural, happy, and also stimulating exercise of the sex-functions, the periods of complete abstinence should be opportunities for transmuting the healthy sex-power into work of every sort."[14] Sounds like Freudian sublimation, but Stopes never cites Freud. She was generous with her citations, so it wasn't dishonesty, and it wasn't likely ignorance either: Stopes earned her Ph.D. in Germany and was the first woman appointed to the science staff at the University of Manchester in 1904. Rather, Stopes and Freud were both operating in a world where this metaphor, of a limited amount of energy, dominated the way people thought.

Freud's concept of libido was a person's sexual energy. It was limited, to the extent that if you didn't use it for sex, it became available for art. But Freud always proposed that this libido could increase and decrease, and by 1925 he had turned away from the whole idea of libido as an actual energy. In the 1920s and 1930s, one of the major things the culture took from Freud's work was that mental discomforts were in part the result of repressed sexual desires. The body and the mind had to get in balance, but this discussion was no longer at all about how much energy people had to spend. Unconscious desires should be made conscious and perhaps acted upon. Freud mostly took individual happiness as his primary responsibility—social conventions be damned—and his ideas led directly into the sexual revolution later in the century. It was Freud's student psychologist Wilhelm Reich who announced the revolution. Reich's book *Character Analysis* (1925) had a revolutionary impact on psychoanalysis itself, shifting attention from symptoms to the whole character. The book also argued that unreleased psychosexual energy produced physical blockages in a person's muscles and organs, as well as blockages of character, and that these acted as a "body armor," preventing the release of further energy. One way to break down this armor was to orgasm. It is another fascinating and inventive metaphor. It

did not catch on. Yet these ideas grew into a broad theory of the importance of a good sex life. Reich coined the phrase "the sexual revolution."
His book of that title came out in 1936 and argued that if you want to be
happy, you need to release repressed sexual tension. Not only should we
explore our desires; we must hunt them out, rouse and enflame them.
He was much too attached to the notion that this energy was a real
thing. In fact, a court found his therapy fraudulent and forbade him
from doing it; he defied the rule, served a year of his sentence, and died
in prison. It can be very dangerous to find yourself alone in yesterday's
metaphor, enthralled in a trance of value that now seems ridiculous to
the point of fraud.

The modern sexologist Alfred Kinsey counseled that a happy life
needed sex. For his bestselling books *Sexual Behavior in the Human
Male*, published in 1948, and *Sexual Behavior in the Human Female*,
published in 1953, Kinsey and his assistants asked people how many sexual partners they had had, whether they engaged in oral sex, group sex,
or homosexual sex, and other rather personal questions. His conclusions
surprised people: apparently, there was a lot more of all this going on
than people had guessed. As with Ellis, Stopes, Freud, and Reich, Kinsey
was angry about the sexual repression of the world in which he had been
raised. His results also demonstrated that many people felt injured and
frustrated by sexual conventions. Over the next decades, a lot of the weird
sexual rules of the previous century were looked at anew and discarded.

Of course, the sexual revolution of the 1960s and 1970s was also
brought into being by penicillin and the birth-control pill. These two
marvels of modern science removed, for a while, the most serious traditional threats of sex: deadly disease and unwanted pregnancy. The old
electrical-switchboard idea where a person could take only so much,
and was often better off not taking any, no longer seemed true. More
persuasive now was the idea that your life had to be kept in balance, but
the more you used it, the bigger it got, and it could grow to any size.
That went for your life and your character, and it was conceptually reinforced by the phenomenal growth of the economy in post–World War II
America. Gone was the image of the body, and the economy, as a physical machine of set proportions—an always shifting system that, nevertheless, has to fit back into the same box at the end of every day. Not

even an animal metaphor seemed right, because that would imply reaching a maximum size at some point. The new idea of the economy, individual character, and a healthy sex life seemed like a forest or perhaps a stew pot. Too much of some quality could ruin everything else, but as long as things are balanced and keep moving, in and out, there are no real limitations.

Some say the sexual revolution was a lot of fun. By other accounts (or even the same memory at another time), it wasn't much fun at all: women felt coerced into sex, worried that if they refused they would seem conservative or repressed. For men and women, there was the question of living up to the new freedom and all the pleasure they were supposed to be having. I hope if anyone is squeamish about language they will find this poem worth its little shock. It is by the English poet Philip Larkin, from his book *High Windows*, first published in 1974. I offer it here as evidence that it matters when you live. It is also evidence that it does not matter, but if we take it at its word, the poem believes it matters when you live.

High Windows
When I see a couple of kids
And guess he's fucking her and she's
Taking pills or wearing a diaphragm,
I know this is paradise

Everyone old has dreamed of all their lives—
Bonds and gestures pushed to one side
Like an outdated combine harvester,
And everyone young going down the long slide

To happiness, endlessly. I wonder if
Anyone looked at me, forty years back,
And thought, *That'll be the life*;
No God any more, or sweating in the dark

About hell and that, or having to hide
What you think of the priest. He

And his lot will all go down the long slide
Like free bloody birds. And immediately

Rather than words comes the thought of high windows:
The sun-comprehending glass,
And beyond it, the deep blue air, that shows
Nothing, and is nowhere, and is endless.

The poem has a definite sour charm. Larkin lived from 1922 to 1985 and reported that history, in his lifetime, seemed marked by a progression of enviable releases from social bondage, only some of which had arrived in time for him. It harbors envy and anger; imagining the joy of "going down the long slide" (a splendid invocation of ease and also of sexuality); then a reverse envy where he counts his own blessings; and then that transcendent last stanza, where he takes himself (with some coaxing) into a place beyond these calculations. Talk about "through the looking glass." It feels as if in trying to escape the mundane, Larkin gets momentarily stuck in the windowpane itself, and then pushes through, out beyond beyond.

For those who took a more active part in the sexual revolution, there were other kinds of trouble. Most important, both men and women got their hearts broken. What kills free love is not convention; what kills free love is romantic love. And then came AIDS. By the 1980s, fear of the spread of HIV effectively ended the sexual revolution. Sex is great, but if you find something else to do tonight, you won't have to get tested. AIDS is a big part of what darkened the image of the sexual adventurer.

Yet despite the darkening of public rhetoric about sexual freedom, there has developed a huge, underarticulated assumption that a happy, fulfilled life must have some regular amount of sex in it. Sitcoms and TV dramas often have a character mention not having sex in a while— say, seven months—and the surrounding characters respond with a scolding insistence, jokes, and phone numbers of possible partners. We know what we are supposed to be doing. The other half of that historically odd social insistence that you should be having sex is that you should be having it in an essentially monogamous way that doesn't much call attention to itself.

TV characters who have sex with many different partners are today depicted as a bit sad. It has not always been this way. Consider a few of television's sexual profligates. The sexual freedom touted in the free-love doctrine of the 1960s made it to the small screen in 1970s cop shows, comedies, and melodramas. Fonzie, on *Happy Days*, was representative. The carnal encounters are implied: we see the sexual virtuoso summon a woman to his arms, and in a later scene we see her kissed good-bye. The assumption is always that everyone had a good time: the man is a hero and the woman enjoys him in this special category. As the culture presented it, there wasn't anything wrong with John F. Kennedy, and there wasn't anything wrong with the Fonz. The men had to be a bit sly so as not to hurt the women they juggled, but they were sly, so nobody seemed to get hurt. Fonzie was happy. This view of sex play as heroic and congruent with happiness did not last.

On the 1980s sitcom *Cheers*, the Sam Malone character is a playboy bachelor. Only the show's hypereducated characters, like eternal graduate student Diane and the fussy psychologist, Frasier, seriously question Sam's behavior and what it means about his happiness. Most other characters assume he is happier than they. As with Fonzie, Sam's conduct does not cause him constant trouble. The women he sleeps with often have built-in reasons not to be annoyed by his behavior: a favored motif is the stewardess, in town only for a layover and so complicit in the brevity of their encounter. They come to him for what he is good at; they are women who want to have fun. Also, Sam sustains friendships with men and women, and is admired and respected as a group leader. Critical Diane argues that Sam's behavior is a symptom of psychological disorder and plain old sadness, but she is depicted as an unreliable witness, too unhappy herself to be taken as a guide. Who is living a rich life, Sam, or his critics Diane and Frasier? Sam, by a long shot. Even these critics love and admire him; he has a life full of friendships and encounters, while they gripe of being impoverished in one way or another. In the world of the show, Sam is allowed to be happy.

In the 1990s *Cheers* spin-off, *Frasier*, radio-show producer Roz is a sexual player on the order of Fonzie and Sam, with few differences because she is a woman. She seems to want sex for fun and happiness; she likes it, and likes men. She is shown having a great time flirting, and she reports euphoric sexual encounters. But Roz is also often shown as disappointed. She is chided by Frasier and his brother, Niles, also an effete psychiatrist, and they often pathologize her behavior. She herself is a little sad, but she does not join in on their critique of her. She sees their sexual behavior as prissy and overthought. Even they cannot help admiring her sexual virtuoso skills now and again. It is interesting that Sam's character from the show *Cheers* visits the *Frasier* show in 1995, and now he is deeply ensconced in a new identification as a "sexual compulsive," with a therapy group and a commitment to change. Now too much sex is a problem that betrays unhappiness and requires the intervention of the medical community.

On the TV comedy *Two and a Half Men*, of the following decade, the many-partnered Charlie has a great deal of fun. His happiness is announced, explicitly, in almost every episode. Yet with the same frequency, Charlie and his family and friends all note that his hypersexuality is an emotional impairment—almost a psychological disorder. He pleases women a lot at first, but many of them end up hating him. They do not seem to have short-term desires for sex with a good-looking, successful playboy, and are often hurt when they find out this is all that is on offer. His seductions are more canned than the women realize, and when they find out, they are hurt. Yet in some ways Charlie is a particularly sympathetic character. For all Sam's honesty, we never really see why he rejects romantic love, whereas with Charlie, his mother's narcissism and overt sexuality are regularly on display. The eyes of a child don't lie, and the half-a-man, the ten-year-old Jake, sees his sexual uncle as happier than his tightly wound, often celibate dad. But unlike Fonzie and Sam, who had a whole clutch of people who were unabashedly envious and impressed, Charlie gets only passing praise of his prowess. The only adult in Charlie's life who supports his sexual behavior is his gruff, nasty cleaning woman, who likes it for its lack of sentimentality, though she, too, commonly treats it as a form of emotional retardation. There is a regular character, Rose, who represents the women who have been harmed by Charlie's sexual virtuosity. Charlie himself is often brought to announce that he is not winning at sex but failing at intimacy.

It seems to me this trip down the history of television Don and Doña Juans reflects a remarkable shift in the culture. Everything outside the private sexuality of romantic love is increasingly depicted as a species of sadness or mental disturbance. Again, these are not laws or sermons, and if you go against them, you won't lose your friends, get fired, or be arrest-

ed. Still, we know what we are not supposed to be doing. It is remarkable that these television sex hounds are not being warned and scolded
about AIDS. Instead, they face this idea that to be fulfilled and happy,
you should be with one person at a time, and not in any way that would
make a great story. We judge people happy who have a relationship
with someone and about whose sex lives, after the first blush of early
romance, we don't hear anything at all.

We might want to say that we have more allowance for homosexuality
than has any culture before, but I think we need to be cautious. Gay
and lesbian people have more "rights" than they have had through most
of recorded history, and one hears many expressions of tolerance, but
there have been many periods when people barely noticed homosexuality as something to have an opinion about. Some people have encounters, or set up house, with people of the same sex. Yeah, and some
people like scary movies. The question is, do we notice it as a thing to
judge? Here is a strange example of what I mean: In the European Middle Ages, over the course of several centuries, there was a cultural bias
in favor of fish over meat. The richer you were, the more often you ate
fish. Over these centuries, there were many recipes for making animal
meat feel and taste like various kinds of fish. There were all sorts of
debates about the best ways to do this, and nary a whisper about why
everyone was so sure fish was better than meat. They must have thought
that someday there might be a paradise where we could all eat only fish
all the time, but that is not how their problem was going to be solved by
history. Now we love steak. How to make it taste like flounder is really
out of the area of concern today. We don't care about who tongue kisses
and who doesn't enough to encourage the tongue kissers to mark out
enclaves for themselves in the larger cities.

How much sex are we all actually having? In the early 1990s the University of Chicago did a sex study of Americans, which today stands as
the up-to-date version of Kinsey's reports. It was done with face-to-face
interviews of a random sample of close to thirty-five hundred people,
aged eighteen to fifty-nine, using sophisticated modern polling techniques. Sociologist Edward Laumann led the research team. The study
shows the frequency of sex in America (with another person) to be as follows: 14 percent of men surveyed had no sex in the past year; 16 percent
had sex a few times over the year; almost 40 percent had sex a few times

a month in the past year; 26 percent had sex two or three times a week; and 8 percent had sex four or more times a week. The numbers for women are almost exactly the same (10, 18, 36, 30, and 7 percent, respectively). Married couples have the most sex: close to 40 percent of married people say they have sex twice a week, compared with 25 percent for singles; and married couples are the most likely to have orgasms when they do.[15] The 5 percent of men with five or more partners do have sex more often than men with one partner, but in all other cases, the more partners you have, the less sex you have.

These stats show a different world than we see on television and movies. For one thing, in a year, 30 percent of men have either no sex or sex a few times. From the media, we might have thought sex at least, say, twelve times a year was more general. The media also suggest that the guy who sleeps with a few different women over the course of the year is a playboy and is having more sex than the average married couple. But no, it is the monogamous, cohabiting couples that need a hose turned on them. Doesn't that suggest that every married person is so because they made a calculation and decided this was the best way to have a lot of sex, and the person with many partners is willing to forgo sex in deference to some other value, perhaps the studious acquisition of curious tastes? The playboy seeks challenge, variety, and (sentimental) education, while the married stay home and rut.

Consider one last note on the historical nature of sex. The researchers of the sex study I have been discussing published a book in 1994 detailing the results and what they thought it all meant, and tellingly called it *The Social Organization of Sexuality.* Their close study led them to believe that the sex lives of various Americans—grouped by gender, age, class, and ethnicity—were all heavily determined by social rewards and pressures.[16] Cultural changes, they found, alter sexual practices a great deal. For instance, "The proportion of men experiencing oral sex in their lifetime increases from 62 percent of those born between 1933 and 1937 to 90 percent of those born between 1948 and 1952. The significant increase in lifetime experience occurs for cohorts coming of sexual age (twenty to twenty-four) just prior to 1968."[17] Where Kinsey saw the low numbers of oral sex as a reflection of social constraint, the researchers of this new study see society as creating the expectation of oral sex. The act seems to have become more mainstream on college campuses in the 1960s, and to have remained more prevalent among the well educated. "The current prevalence of oral sex in the United States is the product of a sociohistorically based shift away from the traditional script of the sexual event between women and men."[18] So that's why you do it—and here I bet you thought it was your idea. Indeed, get this: we can predict how much you and your partner are having sex based on three things: your age, how long you have been together, and whether you are living together (either married or shacking up); these are the only things that help predict your frequency of sex. Race, religion, class—none of it means anything. Everyone, across the board, who has been married twenty years and is fifty years old is having about the same amount of sex, whether you are a rich blond religious woman in Georgia or a secular African-American professor in Massachusetts. What race, religion, and class can predict, by contrast, is what you do in bed. Professionals and "white" people engage in varieties of sodomy much more than people in working-class and African-American communities. It is fascinating that cultural myths do not control frequency of intercourse but the myths do control the rest of what goes on in bed.

The 2005 biannual study "Sex on TV 4," by the Kaiser Family Foundation, found that the number of sexual scenes on television has nearly doubled since 1998. The study examined a representative sample of more than one thousand hours of programming including all genres other than daily newscasts, sports, and kids' shows. All sexual content was measured, including talk about sex and sexual behavior. The study found that 70 percent of all shows include some sexual content, at about five sexual scenes per hour, compared with 56 percent and three scenes per hour in 1998. Guess what percentage of movie scenes include some sexual content. The study put it at 92! Sitcom scenes were sexy 87 percent of the time; 87 percent of drama-series scenes; and soap operas came in at 85 percent. The genre with the lowest sexual content was "reality shows"; it is clear that reality has less sexual content than television shows do. On reality shows, sex shows up in 28 percent of the scenes. It is enlightening to see how much more stamina we seem to have for watching a show about sex, as compared to the real thing. This seems to be something we have accepted. It would be hard to imagine many people's real lives keeping pace with the amount of sex and sex talk in movies or sitcoms.

Note how strict our cultural judgments are, even while we tell everyone to do their own thing. Perhaps you are married, and you have sex a few times a week or a few times a month. You may judge your sex life as satisfying but conceptually dull, the ultimate in normal. I invite you to notice that our version of moderation is just another wacky regimen. Note too that only a bit more than half the relevant population fits this "norm." Again, my chief purpose is to reveal that our beliefs about what we ought to be doing are far too heavy-handed, and nothing reveals this better than a little historical perspective. Consider the absolute conviction that people had about a handful of outlandish ideas, based on changeable arguments, and almost no real attempt at proof. It should wake us to a more realistic understanding of our own convictions.

All this said, I think our era allows more room for happiness through sexuality than average. There are constraints of our time period. There is AIDS. There is the fact that, as a result of our long educations and tight economy, the biological readiness for sex comes much earlier than the social and economic readiness for its consequences. There is also the fixation with homosexuality as problematic. And there is a kind of boring silence about sex, because when it becomes visible, people interpret this as a psychological "issue," which may or may not be fair. Nevertheless, you can have a pretty good time here in the twenty-first century. Past centuries have been better with allowing people to try for happiness by using drugs or through an economy that values craft, community, and leisure. We're pretty good with sex. There is room to move, without having to change any laws or overhaul entire institutions. Of course, you do need a little historical perspective.

16

Treatments

You are lying on the beach in a quiet cove in the Yucatán, listening to softly crashing ocean waves. When you open your eyes, you see palm trees on either side of you; behind you, in your stilted cabin, your sun-shy spouse sleeps in. You can see the narrow beach on either side of the cove extending emptily far into the distance. You check under your bathing suit to see how dark you are getting, then feeling a little guilty for wasting the opportunity to go into the ocean, you haul yourself up and take a dip in the turquoise water. The waves are rough enough to keep you busy. Still, you feel a sharp pang of bliss as you float and gambol, letting the ocean and the sky press you back and forth like two kids who won't let go of a ball. One wave surprises you, though, and knocks you down; you clamber out and collapse back into your spot on the sand. You think of the office you work at back in the States and of how long you looked forward to this trip. You sigh with pleasure, lick your salty lips, and close your eyes. Happy.

Just what are you happy about? It is not easy to say. Many people report that the sun feels good on their skin. It seems like an objective truth when you see how crowded Jones Beach and the Santa Monica Pier get on a bright day. But what about the people of Mexico and Jamaica, for instance? They do not seem to find their sunny beaches nearly as compelling as their tourists do.

Natural beauty and quiet are both a delectable change from strip malls, television, traffic, copy machines, chattering, and cubicles. It isn't exactly interesting, though, and the ocean isn't exactly quiet. It is loud and monotonous. You know how nice it is when you are resting and someone shuts off the lawn mower or the vacuum they were using? At the ocean side, if there were a button we could press to make the surf stop for a while, a lot of us would probably press it. As for the natural beauty, note that the palm tree has become a symbol of happiness. There was an obvious reason to correlate the tree with warmth and leisure, but by now it has taken off on its own, standing as such a strong symbol of paradise that a beach and sun without palm trees would be like a birthday party without balloons. That is, nowadays, the palm trees are doing some of the work of making you happy. What else? You are alone. Part of our current fantasy of a happy vacation is that you have the beach to yourself. Note again that this version of happiness goes against common desires to go where all the people are. How about that experience in the ocean? Euphoric, and a lot of work.

It is all great, but it is also a specific trancelike fantasy that allows us to go to a "paradise" while on vacation. If every year you take three months off, the pleasure of their passage can be mild and diffuse, but if you get only a week, you'll need some strong symbols of delight. I don't mean to say these vacations do not just plain feel good in a way apparent to most people, regardless of their culture. Some parts of our vacations are essentially the same as what the ancient Romans did to relax and get happy. They didn't care at all about palm trees, but there are some remarkable continuities. Some of what we enjoy on vacation is just the resting that we allow ourselves to do there. Might some such symbols help us authorize ourselves to rest for days on end at home?

Along with the modern myth of the palm tree, consider the image of the orgasmic shower. The Clairol Herbal Essence shampoo commercial was memorable and remains familiar: a woman moaning with pleasure as she takes an ordinary shower. Companies make body washes, shower, sponges, and other paraphernalia and advertise them as luxurious, euphoric, and dreamy. Why are we imagining ourselves having such a good time in the shower? We don't traditionally love washing things. We are on our feet in there, so not exactly resting. What sense does it make?

These two modern images of happiness are part of a long history of treatments and practices that promise bliss and relief. There are water cures, mud baths, acupuncture, yoga, chiropractics, high colonics, saunas, steam rooms, tai chi, sitting meditation, ecstatic dance, Swedish massage, and shiatsu. There is also bleeding, cupping, purging, and flushing the system. This latter group was based on the doctrine of the four humors. It is an old one. Hippocrates, our earliest historical doctor (he lived from 460 to 370 B.C.E.), already knew about it, but it was our second historical doctor, Galen (131–201 C.E.), who first famously advanced the idea. The four major fluids in the human body were happy blood, groggy phlegm, anger's yellow bile, and the black bile of depression. *Sanguine* means "bloody," and since blood was the happy humor, *sanguine* means "merry."[1] It wasn't that blood was happiness itself; it was that happiness was based on having the right balance of blood. Thus, misery could be treated by leeches. Anxious people, in particular, seemed to need a bit of a draining to be happy. The doctrine had influence across millennia, and was not fully dismissed from medicine until the nineteenth century.

Until then, drawing from the idea of the four humors, doctors took blood and induced their patients to vomit, sweat, or excrete.[2] Sometimes the doctor rebalanced the harmony of a patient's humors; at other times the patient died—from the original complaint or the treatment. Since these treatments persisted for so many centuries, though, it makes sense to guess that they worked, in some way, and to investigate how. Of course, often the cure did little harm or good. Sometimes the methods helped because painful treatments bring endorphins or even an adrenaline high. I think here of trepanning, acupuncture, and extremely strong massage. Being bled makes one lightheaded, and, of course, lowers blood pressure. Also, there were potent medicines that were unrecognized aspects of adjusting someone's humors, from analgesics given for the procedure to the anticoagulant properties of leech saliva, which can travel throughout the human circulatory system, dissipating blood clots. What we call "heroic medicine" tried to heal people by bleeding, sweating, purging, and blistering them; one justification for such treatments was that you could hasten the symptoms along by bringing them on. The methods could be especially potent if the problem was psychological. Listen to Montaigne in 1580: "How many men have been made sick by the mere power of imagination? We see them regularly, having themselves bled, purged, and physicked, to cure ills that they feel only in their mind. When real evil fails us, knowledge lends us hers."[3] Most everyone who offers a happiness treatment in today's America has a rationalist theory of why their treatment works. People don't announce that they are poking you with magic needles. But just being touched by attentive hands helps a lot of what ails many of us, and the ritual of being taken care of can be affecting, whether it takes place in the theater of medicine or under the lavender lights of a spa treatment room. There are grooming treatments that provide comparable effects. This is part of the reason there are so many nail salons, and why so many hairstylists include some massage, hot towels, and fragrant elixirs. Barbers used to be surgeons, of course, and not only because they were the only ones in town with a straight razor. Grooming shops and medicine are all of a family: providers of happiness treatments for the body. They are where you go to get taken care of.

The great historical happiness treatment is the water cure. When building a new town, the public bathhouse was the first thing the ancient Romans usually constructed. It was the Romans who made the Belgian town of Spa famous for its spring waters, which is where we get our word. Spas were an opportunity for social travel, going someplace else: high or low altitudes, clean air, dry or moist climates. Doctors prescribed retreat to such climates, and as a result, sanatoriums arose there. In history, the most enthusiastic spa-goers were the sophisticated ancient Romans and the urbane cosmopolitan bourgeoisie of the European Industrial Revolution. Spas are rural outposts for a tired urban population that sees itself as so depleted that it needs to soak in minerals.

Roman spas were fairly routinized: you arrived, got naked, had oil rubbed on you, then went into a warm room full of couches where people would lie around talking. Later you'd move on to a very hot bath, where you would sit and sweat, scraping your skin with a special tool. Often they had slaves to do the scraping.[4] Then it was a quick dip in hot water, then a quick dip in cold water. In 70 C.E. the Romans built a spa around the hot springs at Bath, England, that bubbled magnesium, potassium, sulfur, and calcium. By the year 300 C.E. there were over nine hundred baths throughout the empire, and at the height of this water indulgence, thirteen aqueducts supplied ancient Rome with three hundred gallons of water per person per day. There is no greater act of conspicuous consumption than letting fountains spill water down parched streets. Soaking is, in part, a display of having enough water.

At various baths across history, there have been treatments that assault the patients with water. The logic behind this is that human bodies are bags of various fluids—call them humors or hormones or chemicals—and vitality is maintained through the movement of these fluids from their points of origin to the rest of the body and then either out or back to where they started. When the liquid needs to move down, gravity does it. To move liquid up and around, we have the pumping heart, and our voluntary movement, which squishes and squeezes everything around. The more sedentary we are, the more reasonable it is to assume that our juices will get stuck—amid muscle fibers, in the extremities, and in the gut. Western medicine speaks of moving lymph through the body; Eastern medical arts speak of *chi*. Tai chi is a meditation you do while moving, so your chi can flow. Similar explanations are offered for the pleasure of being massaged, acupunctured, sweated, and helped into strained positions, from yoga to Pilates. All these behaviors have been reinvented many times since the ancient world, often with startling enthusiasm. In the mid nineteenth century, James Caleb Jackson presented water therapy as so stunningly new that it had opened a breach in history. Wrote Jackson, "The water-cure revolution is a great revolution. It touches more interests than any revolution since the days of Jesus Christ."[5]

In the twentieth century, the bubble bath became a symbol of pampered female life. Odd, because a bubble bath costs almost nothing. What makes the bath seem luxurious is having the time to take it, and what makes it seem pampering is that (for many people) it works: it makes you feel good. But the suds themselves seemed to hold the luxury, as if the foam were a white fur stole around bare shoulders. The white bath is for women, and the leisure it symbolizes is provided by a man's wealth. Especially in the movies. It may be that the original cinematic bubble bath was not a bubble bath at all. In *The Sign of the Cross* (1932), Claudette Colbert, as Nero's empress, swims around in a lavish pool-sized bathtub filled with donkey's milk (we are shown the many teams of mares and milkers), and as she cavorts through this pre-code party, she allows glimpses of her naked breasts, lapped by the tide, and orders another pretty woman to strip down and join her. The Hayes

Code stopped such things by 1934. The visual memory of all this does not leave one quickly. Especially when, after you invent it and sell it, they make a law that no one is allowed to do it again. It seems the bubble bath, another white cloud of liquid, maintained the milk's extravagant luxury. In the film *The Women* (1939), Joan Crawford's Crystal is a sexy, savvy, bitchy husband-thief ("You noble wives and mothers bore the brains out of me") and practically lives in her extravagant Art Deco bathroom. The tub has carved waves on the sides, two satin pillows for a backrest, a stand for her magazine, and ample surfaces to support a phone, a mirror, and an ashtray. Her bath is always frothy with bubbles, so people visit her in there—a maid, the young daughter of the first wife, a society friend—and all trade catty remarks. Chided that she's been in there too long, Crystal lies back, a smoldering cigarette in one hand, a bitten chocolate candy in the other, and barks, "The doctor ordered me to soak in this foam for my nerves." *Love Moods* (1952) features burlesque star Lili St. Cyr performing her celebrated bubble-bath routine. The bubbles covered her girl parts. By the time *Some Like It Hot* came out (1959), the bath as the ultimate in femininity and luxury was a bit of a joke: Tony Curtis uses a bubble bath to hide that he has boy parts. He had been down at the beach seducing Marylyn Monroe's Sugar, using information he had gleaned in drag as her girl bunkmate. The idea that the bath is more than a slow way to cleanliness continues on through the history of film, revealing the preoccupations of the various decades.

Let's zero in on a little film from a period trying to hold on to the postwar dream of family while beginning to deal with the fact that women wanted a better role in this bountiful new world. The 1963 film *The Thrill of It All*, written by Carl Reiner and directed by Norman Jewison, tells its story in three bubble baths. In the first bath scene, Doris Day, as Beverly, gives her daughter a bubble bath while her small, squeaky son looks on. The girl demands use of mother's new soap, mother relents, and the nice-smelling soap makes the girl happy. Beverly is stuck in the bathroom on her knees this whole long scene, while her kids and her husband (on the phone) markedly annoy and exasperate her. Next, Beverly and her husband, James Garner's Gerald, arrive at a dinner party at the Fraleighs' (sounds like "folly"), where Old Tom Fraleigh, his aging sidekick son, and his ad man are all riveted to the television, waiting for a commercial. When it comes, it features a sex kitten of a girl naked in a bubble bath. The girl is daft and luxuriates in the bubbles with starlet camp, cooing: "Find true happiness in your bath. Just you, and a cake of Happy." Beverly mentions that this was the soap brand that "saved her life" that afternoon and tells about the hassle with her daughter. Old Tom Fraleigh announces that his family makes the soap, and the commercials, and that he wants Beverly as their new pitch girl, at a large salary. The rest of the movie is about Gerald hating Beverly's new life. He is an obstetrician, and she was, previously, an ever-available doctor's wife. He was always running out on her (to deliver babies), and now she's always running out on him. But Beverly loves her new work.

The third bath of the film is a doozy. One day, while Beverley and Gerald are out of the house, the Fraleighs take it upon themselves to put a pool in the couple's backyard—in order to film a commercial poolside the next day. In preparation, they leave a large quantity of Happy detergent at the pool's edge. That night, Gerald comes home after dark and drives right into the new pool. Furious, he stomps off to spend the night in a hotel, and as he goes, he knocks all the soap into the pool. Late that night a rainstorm comes in, churning up the soap until huge sudsy clumps assemble. Some float into the yard's tree branches. It is a beautiful scene, dark and surreal. Often a box of soap is visible, showing no logo or ad patter, only the word HAPPY, floating across the dark pool or bearing up under the rain. Beverly wakes up to a blue sky, and a backyard completely full of soap bubbles—a mass of foam bigger than the house. She opens her second-story bedroom window, and the suds pour into the room. When she and the kids survey the mesa of suds in the backyard, there is a fairy-tale or sci-fi look to the image. This encounter with the soap-selling world got out of control fast! It is too much. Beverly cleans up everything. She has the bubbles carted away and has a crane raise her husband's car out of the pool.

Of its three baths, the film pronounces the domestic first bath irritating and frustrating. The second bath was a cheap fraud: a foolish girl pretending to be sexual in order to be an actress. The third bubble bath was absurd, immense, and magical. Its pool of origin was a sign of the wife's growing power in the house, seeded by her husband's anger. How does the marriage survive? Magic. Beverly and Gerald happen to end up in the backseat of a car, together delivering the younger Fraleigh's baby. The experience makes Beverly concede that her husband's work is more important than hers. She vows to quit her job, and they head home to have some lively and explicitly procreative sex. The value of babies versus selling soap is one every family has to negotiate, but usually the woman makes babies and the man sells soap. Theirs is a crazy little fantasy solution to the battle of values, blending male and female imperatives in a way available only to couples where the man is an obstetrician and the woman's career is as dispensable as selling soap is here. There is no question that what we are fighting over is happiness; it says so right on the box. Note that the good wife never herself got into the tub: she bent beside one, watched her husband ogle the starlet wannabe in the soap commercial, and then sort of conjured up the last one at a distance; it was awesome, as bubble baths go, but it didn't quite happen to her.

We may begin to wonder if a good woman ever gets in a bubble bath. Dudley Moore's title character in *Arthur* (1981), a parody of dependent wealth, was emasculated by his bubble baths, but at least he got to get in and relax without being a slut. Julia Roberts's character in *Pretty Woman* (1990) has a bubble bath that carries the traditional bubble-bath symbolism: female luxury and ease provided by a man. Note that this works in the 1990s only if the woman is an actual prostitute, not just a mistress. Just like the good wives of an earlier era, women with productive, legal jobs do not have the time for the bath, nor the need to rehearse the bubble bath's symbols of luxury, sexuality, and being taken care of. It is interesting to contrast the bath Diane Keaton's character takes in *Shoot the Moon* (1982). Her husband has left her for a younger woman, and one night when the kids are out of the house she takes a bath and smokes a joint and sings and cries. That is no frothy bubble bath because not only do bubble baths symbolize happiness, they also symbolize protection, and in that moment what makes sense is the blunt vulnerability of clear water. Movies seem to have a definite idea of what bubble baths mean, and who gets to take them. In real life anyone can take one, and when they do, they get to rehearse the experience of what the bubble bath means: protection, luxury, sexiness, leisure, naughtiness, or even amorality.

Nowadays not many people take bubble baths. They say they don't have the time, but they make time to watch *American Idol*, right? They are not sufficiently interested in rehearsing the bubble-bath symbology.

Advertisers claim their soaps make so much lather that a shower can have the same prestige as the soapy bath. The image of the blissful shower is not about cleanliness. It is about opulence and leisure. By pretending the bubbles were the bliss, instead of the long soak, modern advertisers try to transfer the mythic bliss of the long bath—through the bubbles—to the shower! It works because, though soap is not expensive and leisure is, soap bubbles have come to symbolize the wealth of leisure and can now bring a certain amount of happiness to the ablutory event, even without the leisure. The suds also don't do their job of enclosing the body when you are standing in a shower, so just being able to create them on a sponge is not that gratifying. Still, the shower has its own pleasures. As humanity has noticed over history, water feels nice. In the shower, the noise of the falling water gives us a moment away from family, phone, and e-mail. As we saw in *The Women*, the bubble bath hid your body, so that people came in and talked to you, even fought with you.

The shower means a moment of privacy. Advertisers of shower products should shut up about suds and their attendant meanings of luxury, sexiness, and protection and instead remind shoppers to take showers because they feel good, they drown out all our electric bells and beepers, and you get to be alone. If I wanted to make money on shower products, I'd think of something to make you stay in there longer—tingling after-scrub shower lotion, perhaps—and I'd show a model in the shower daydreaming and absentmindedly applying the product while outside the bathroom door her kids holler, dogs bark, phones ring, and e-mail messages chime their arrival. In the 1970s it might have seemed like a good idea to hurry out of the shower and back to the fun, but these days the fun has gotten ubiquitous and overbearing, and we need a break. You can keep your combination shampoo and conditioner. Soap moguls: don't give them less to do in there; give them more. It is the most peaceful spot in modern life. For many women and men it is also the most private time of their day. Only the car affords them so much privacy, and there the blind but watchful eyes of all the other commuters come in and out of focus with the speed of traffic. Only in the shower is there a total reprieve.

Along with water treatments, spas have often had some kind of program of self-denial built into them. There was indulgence in treatments, but spas were advertised as a place to become healthy, and if you are going to charge someone for something as amorphous as that, you'd bet-ter *do* something to them. It is also true that self-denial works. Often physically and psychologically painful at first, a self-denial routine can come to feel very good—bracing, clear, and bright; it's often described as feeling as though you had "broken through" to a new place. Even to the extent that self-denial stays painful, it can be very distracting, and sometimes people need distraction. Further, in the past and today, many denial regimes come with a lot of community. In a community of a given regime, the regime puts everyone in a similar mood, and you sup-port one another in the catechism of the doctrine. In general life, in any century, it is not at all obvious how to be virtuous. In a dedicated com-munity, it is clearer; indeed, you can be otherwise unproductive, but val-ued for your iconic obedience to the laws.

There are several ways that a culture may embrace self-denial. Here is one: It was about three centuries after Jesus died that we first see evi-dence of anything like a forty-day fast before Easter. In his "Festal Let-ters," St. Athanasius announced that his Alexandrian flock would have to start observing forty days of fasting preliminary to the stricter fast of Holy Week. In 339, having traveled to Rome and much of Europe, he wrote that people were going to have to start taking this seriously "to the end that while all the world is fasting, we who are in Egypt should not become a laughing-stock as the only people who do not fast but take our pleasure in those days."[6] People don't fast only because it reminds them of death and hardship, but also because they want to fit in. Note that at one point this fast meant that you ate only one meal a day, avoiding all animal products. Later that was relaxed so that pious donations could be traded for a bit of dairy, which is why one of the towers at the cathedral at Rouen was long called Butter Tower.[7]

Here is another story of the rise of a culture of self-denial. It takes place in two utopian dreams on the Jersey shore: the "twin cities" of Ocean Grove and Asbury Park.[8] These were vacations spots—visions of happiness. Ocean Grove came first. It was a planned "Christian Seaside Resort," and that meant Methodist and temperance. In the 1880s, its summer population was around twenty-five thousand. Where Ocean Grove met land there was a wooden fence that was locked every midnight and all day Sunday. There were strict edicts against liquor, tobacco, novels, cards, and dancing. To make quite sure no business was done on Sunday, they emptied the penny-candy machines on Saturday night. The authors of the Baedeker guidebook couldn't believe so many people chose to spend their summer vacation under a "religious autocracy."[9] The Grove included a sanatorium "where rest and relaxation may be obtained from the busy turmoils of everyday life in the cities and towns," offering the usual services: special diets, water cures, mineral soaks, internal irrigations, and rub-downs. The center of life at the Grove was, however, the ocean itself. In 1874 Ocean Grove's Reverend Aaron E. Ballard detailed the effect of the waves there:

> The long, rolling surf-waves … gently shock the frame and stir the sluggish blood to fresher motion. That motion rolls, and bonds, and leaps through the veins—anywhere, everywhere—routing all the hosts of peccant humors which have ambushed themselves in all possible hidden places…. The torpid liver finds itself compelled to join the general activity, and to work like a disused steam engine newly set in motion. The nerves respond to the body's school-boy holiday and scatter tingling sensations of pleasure over the frame…. The surf lubricates the joints like oil; grave men fling out their limbs like colts in pasture; dignified women, from the very inspiration of necessity, sport like girls at recess, and aged people tumble among the waves till one would think they were only in their teens.[10]

The mix of the erotic and the idea of the humors is delightful here, as is the industrial-age metaphor of the body as an old steam engine chugging back to life. Most intriguing is that the people depicted are dignified men and women who sport only when forced: the waves take them over, leaving them free to enjoy a good roll.

Ocean Grove's twin city began when James Bradley visited the Grove and, soon after, bought five hundred acres adjacent to it, for Asbury Park. (It was named for the pioneer Methodist bishop in America.) Bradley set up a board of health, an advanced sewage system, and an electric trolley. He announced that "the Park," too, was a temperance town and a Sabbath town. Bradley's tale of why he first went to the Grove, and why he started the Park, was about nervous exhaustion:

Having for some time previous been in bad health, I concluded to try what I had been recommended—sea air. Too close application to business had made inroads on my constitution, and my nervous system was seriously affected.... I have often met persons since the time I first camped out at Ocean Grove whose nerves were shattered by too close applications to their professions, studies or their chase for the "almighty dollar." I was familiar with their sufferings which, alas, strong men look upon with contempt. Some were taking this or that "nervine cure-all," but the best nervine for a man who is not absolutely past repair, is to break away entirely from his calling or greed and camp out on the sea shore ... and patiently wait for the return of ... good health.[11]

Again, I include this long quote because the details are so revealing: the "sea air," the overburdened nerves, the idea that "strong men" look upon such suffering with contempt, the noun *nervine*, the idea of bringing the body to a restful place and passively waiting for happiness.

Asbury Park, too, was very much about community: they had baby parades, pageants "designed for the Puritan more than the Bacchanalian or Bohemian," and they held a "Children's Carnival." At Ocean Grove, fancy clothes were frowned upon, while at Asbury Park everyone dressed up and strode the boardwalk to see and be seen. Writing for the *New York Tribune*, the novelist Stephen Crane noted that people flocked to the boardwalk at Asbury Park "to see the people.... For there is joy to the heart in a crowd. One is in life and of life then. Nothing escapes; the world is going on and one is there to perceive it."[12] It is revealing of our own concerns that we fetishize solitude such that our fantasy is an empty beach. More people live in cities today than in the past, so we are crowded in that sense, and maybe that's why we want to be alone; but due to the housing shortage in those rapidly industrializing cities, today we have more room. We do head to the throng for fun, too, but our preoccupation with beach solitude may be historically unique. Asbury Park was a playful, social, vital place that felt "at the heart" of things: one can there see "the world" going on. Bradley was praised for having built a "city dedicated to health and happiness."[13]

The differences between the two towns were at first minuscule; they were called twins. Both were family friendly and proper, but the differ-

ence was that one changed its mores to follow the middle class, whereas
the other determined to keep faithful to the middle-class preferences from
the time period of its origins. This is a thorny decision that determines the
story of all special projects: keep it going as is, until it dies, or let it change
entirely, and live. After a while, the business and hotel owners of Asbury
Park wrested control from Bradley and introduced liquor licenses and
Sunday amusements. Without the pressure of a population dedicated to
sobriety and an idle Sabbath, capitalism and desire were free to make
alterations to the original idea of the town. From then on, the twin cities
started insulting one another in local newspapers, with the Park imputed
as a pack of sinners and the Grove as a bunch of prudes "who look upon
anything that conduces to cheerfulness as a crime."[14] It is almost comical
how wholesome Asbury Park was—a children's carnival!—but Ocean
Grove had grown more notably about the pleasures of self-denial and the
experience of being part of a community with special virtues. Notice how
these models—happiness by purity and self-denial on the one hand, happiness by fun and self-indulgence on the other—created each other. More
than the profane outside world, it was gently progressive Asbury Park that
pushed Ocean Grove to fall in love with its conservatism. Montaigne
wrote, "The laws take their authority from possession and usage; it is dangerous to trace them back to their birth. They swell and are ennobled as
they roll, like our rivers: follow them uphill to their source, it is just a little
trickle of water."[15] So that is my twin study. It argues that culture counts a
lot, and that cultural mores are really arbitrary. Clearly, every "conservative" culture had its clock stopped at some point. As time goes on, the rendition gets increasingly fictional.

Consider that the marvelous Mohonk Mountain House in New York,
started by identical twins Alfred and Albert Smiley, gracefully rolled
from nineteenth-century respectable fun to twentieth- and now twenty-
first-century respectable fun. You splash in the lake in summer, or visit
on the word-game weekend in winter. Once there was no liquor, and no
visitors on Sundays; now there are both, because those things no longer
mean what they used to mean. Visit, or see the lake and lodge in the
movie The Road to Wellville, where the hotel plays the role of Kellogg's
Battle Creek Spa.

Family spas are not as prominent a vacation destination as they were
a hundred years ago. Yet look at where families do go! Masses of them

spend summers and vacations getting all shook up at amusement parks, on water rides, and in beach waves. Couples go to beaches. Singles go to beaches and spas. Yet we do not seem to be aware that we are jostling ourselves, and our children, to happiness. Convictions about how to live, what rules to follow, are mostly historical accidents of the moment, but we can still see patterns; some underlying factors of various behaviors can be discerned. For instance: get in the water. There are endless historical accounts of peoples' getting in the water making them happy. It is not merely a recreation, nor is it primarily about being clean. It is a happiness technique. Experiment on yourself and see what you think. Get your body shaken up and worked over. The other important consistent message is that self-denial is a potent and common way for people to control their happiness. The removal of something from your menu of pleasures causes acute distress, and then a reawakening, a brightness. The cultural meanings that help to interpret this as happiness vary widely with historical time. Self-denial is reminiscent of virtue, because both take something off your plate. But there is a big difference between limiting one's appetites so as to share with others, and limiting one's appetites in order to be thin and get more attention. Again, they both seem virtuous, but the second one is not. Body discipline may make you happy. But it isn't virtuous. Flexing your willpower is no more meritorious than flexing your muscles: not much. What matters is whether you can step up when it counts.

Celebration

Happiness requires some public processing of grief and fear. More than any other people in history, we are enabled by our culture to feel independent. Our emotional lives are supposed to take place almost entirely within our nuclear family of origin, and in our created nuclear family—bridged by a period where we live our emotional lives safely within a band of friends. Many of us man the borders, keeping "in the family" all information about our private lives. What are private lives? Let's not use the personnel (family and friends) for the definition. Descriptively, I think we can say your private life is where you keep everything disgusting, weak, or wounded; sexual, medical, illegal, and weird; anything involving baby talk, nose rubbing, tantrums. But we really aren't independent; we are a pack—a pack with off-the-charts sophisticated minds and hearts that we use almost entirely to monitor the pack and respond to its every shiver of experience. Sometimes a pack needs to bay at the moon together. Through time and across the globe, some cultures are better at facilitating this than others. We are not great at it, but not terrible, either. What we do manage is often hidden under other names. It takes a little work to see, but once you see it, you won't look at our culture in the same way again.

My chief claim here is that the way we watch the news is not about any rational gathering of information or civic responsibility; rather, it is our way of sharing our fears and griefs. There is geopolitical, economic, social, and scientific news of a sort that affects us all. But there is also a tremendous amount of news that is just the rehearsing of feared intimate dramas writ large: fights, losses, brutality. The way we use these repeated and repeated stories as communal performance of grief becomes most visible and most effective when people attend vigils or get involved in similar ways, and these televised vigils and events feed back into the common experience.

We first-world moderns are not like everybody else. Historically, the average person expected to be a little miserable most of the time, and ecstatic on festival days. We now expect to be happy all the time, but never riotously so. Does the attempt to be happy all the time imply a real loss of the more extreme ends of the emotional spectrum? As a generalization, yes. Agony and ecstasy are linked. Daily comfort does not inspire holiday bursts of jubilation; daily trouble does. As the comedian Eddy Izzard has joked, you'll never hear a more mournful version of "Hallelujah" as when it is sung by an upper-middle-class church; meanwhile, the poorest community, singing the same song, raises the roof with it. Communities that suffer together party together; and community partying is a very different life experience than partying in little hand-picked groups; or, to put it another way, partying within your private life. Partying with a broad community is especially powerful when people know that the party makes possible the sustenance and betterment of the group. This may be imagined as magical but need not be: enactment creates reality. On a private level we understand this. Thanksgiving with the family is not just something an objective group called "your family" does every November. In reality, the definition of "your family" is the people with whom you have Thanksgiving. Showing up at public parties is a similarly foundational act. Taking part in public worry and grief over a news story creates and reinforces the "public private life."

The way this works for us today is most dramatically illuminated by looking at the way it worked in ancient Greek festivals and at medieval Europe's carnivals. Finding out what actually went on at these huge, raucous parties brings a very strong light to what it is that we do today. This section begins with a look at ancient Greeks abandoning themselves in wine, dance, sex, violence, and some weird behavior you didn't know was possible.

17

Greek Festival

Public madness and jubilation in history has been male and female, and has played a similar role for men and women. Still, it is more often performed by women. From Middle Eastern dancers spinning to holy ecstasy to the gospel-singing, hands-waving jubilation at an African American Southern Baptist church, a lot of public ecstasy is heavily female.

The ideal citizen of the Greek democracy was male. The ideal celebrant was female. This has not been pointed out as such, to my knowledge, but my claim is by no means controversial. Women wail and whirl. Nor is it controversial that, as much as the democratic assembly kept Athens great, the festivals kept Athens great, and made her what she was. Democracy was male. Festival was female. So is the enacted madness of mourning in these same cultures. Public ecstasy is heavily poor, too, and heavily powerless. The meek inherit the bacchanal.

In the record of history, the use of the term *ecstasy* is not common. People speak of religious ecstasy, drug ecstasy, sex ecstasy, and ancient festival ecstasy. That's about it. The high at the festival was created by excitements of religion, drugs, and eroticism, but it was mostly about the crowd, and the stories it acted out. Across centuries upon centuries, there was one particular story that kept appearing at the heart of the Greek public ecstasies. This story is worth a close look, not just because it is a lush tale of sex, betrayal, violence, and unfathomably unlikely reunions. It will keep reappearing throughout our history and will shed light on the topsy-turvy of medieval carnival and on every aspect of our own times.

Walter Burkert, a scholar of Greek history, wrote of Greek religion being experienced as "frenzy" and ecstasy: "ritual and institutionalized, collective frenzy, especially the frenzy of the women of a city as they break out at the festival of license."[1] Again, it is women who feature most prominently. Yet there were parties during which everyone in town, regardless of gender, got drunk—for instance, the Wine Jugs Day of Anthesteria. By custom you couldn't break open the earthenware jugs of autumn wine until a particular date in the spring, and Anthesteria was the festival for it. All men and women, including slaves, were each given a jug of strong wine, and whoever finished first won. Children of three years and older joined the fun, with minijugs and real wine. The Greek saying was "Birth, wine jugs, college, and marriage."[2] When infants died they were buried with the minijug that would have been theirs at Wine Jugs Day, had they lived to see it. We have vase paintings of the whole population drunk and wavering down the roads. As one historian put it, "Participation in the time of license creates community and gives the children in particular a new status; the Athenian becomes conscious of his Athenian-ness by the fact that he participates in the *Anthesteria* celebrations."[3] Solidarity must be enacted.

We are going to follow a Greek woman to one of her annual celebrations—the Thesmophoria—and see what she does there. In order to do this, we have to know something of who she is, of what she's walking away from when she heads to the festival. Women obeyed their fathers, and then their husbands. Young women were generally married off to older, established men. The girls kept house, served men, did the weaving, and had babies. Many were closer in age to their eldest children than to their husbands. Though their lives were all about bearing children, they did not have rights to the children. Children belonged to the husband and at a young age would likely be taken away for marriage or war. Of course, children taken away for marriage or war most often visit their mothers, so that the cruel loss also occasioned joyous reunions, and dreams of reunions. By the height of the Greek classical age, it was shameful for a woman's name to be mentioned, even in praise of her obedience and modesty. Women were commonly understood as inferior beings; they were relatively ignored by art, which favored the male nude, and by politics, which categorically dismissed them. Even ideal love was understood as a romance between a grown man and a male youth. The Greek woman was not supposed to be seen outside the house. For rich women this meant life in a lavish courtyard and home, but for most it was sadder even than that. Aristotle complained of how the women of the poor were often in sight, because they had to go to market and do other outdoor chores. We need to know all this because the burdens of these women's lives will dictate the way that they party. Festival is release, the same way you release a slingshot. We can expect an extreme reaction when a slingshot has been pulled far back. We could also say that the further into a particular cultural fetish you take people, the more the culture shows interest in acting out and discussing the other extreme. At Brazilian carnival and at American frat-house costume parties, tough guys dress up as buxom women. As girls, the boys overdo it. Festival brings out strong opposites and makes visible what was most hidden in daily life—here the girl in the man. At the Thesmophoria, women of ancient Greece were suddenly very visible—to each other— running around the woods loud and naked.

For many years I have thought about public ecstasy. I have wondered about what does this work for people today, or if this work simply didn't get done for us. I began by looking at what it was that they had and was surprised to find that for over a millennium the same myth kept turning up behind every wild super-freak festival, the myth of Demeter. The fact that this one story keeps coming up at every hint of ecstatic festival, across over a thousand years, is the kind of surprise that demands investigation. As I tell the story, keep in mind that at many of these ecstatic festivals, people dressed in costume as the various players and at different times acted out the various roles. Remember also that most of these festivals were entirely female.

Demeter was the goddess of grain and the harvest, a sister of Zeus and Hades. She had a daughter called Kore, *the maiden,* a pretty young goddess who liked to pick flowers and play in the fields. Demeter had Kore by Zeus, her brother. We take the incest in Greek myth rather for granted, partly because the gods were all family, but why write it that way? Of course, the problem is worse than the fact that mother and father are siblings. One day little Kore disappears. Demeter goes looking for her, begging any possible witnesses to tell her what happened. The sun speaks up. He has seen what happened to Kore, and, softened by Demeter's maternal grief, he tells her: Zeus allowed his brother Hades, Lord of the Underworld, to take the girl. Hades had used a flower as a lure and caught her up in his carriage. She screamed as he took her away:

> He caught her up reluctant on his golden car and bore her away lamenting. Then she cried out shrilly with her voice, calling upon her father, the Son of Cronos, who is most high and excellent. But no one, either of the deathless gods or of mortal men, heard her voice, nor yet the olive-trees bearing rich fruit ... she cried to her father, the Son of Cronos. But he was sitting ... in his temple ... and receiving sweet offerings from mortal men. So [Hades] was bearing her away by leave of Zeus on his immortal chariot—his own brother's child and all unwilling.[4]

Having listened to the sun, Demeter is in an apoplectic fury at what these men, her brothers, have done to her little girl. She responds by not letting anything grow on earth. She is the goddess of grain and agriculture, and she shuts herself down and it all freezes over. Turning her back on the world of the gods, she dips out of sight, masquerading as a baby nurse for a mortal family—the king and queen of Eleusis—who sense she's a bit of a goddess and minister to her sadness. A woman named Iambe tells stories, cracks jokes, and even strips naked in front of Demeter, until at last she forgets her misery for a moment and laughs.

Faced with a world where nothing will grow, Zeus confronts Hades and makes him agree to return the girl to her mother. There is a rapturous reunion. But as soon as mother and child have thrown their arms around each other, Demeter has a bad thought and asks her daughter sharply, *Did you eat anything while down there?* If she has done so, Demeter explains, the girl will be bound to Hades for a third of every year. The girl responds that when she found out she was allowed to go home, "I sprang up at once for joy; but he secretly put in my mouth sweet food, a pomegranate seed, and forced me to taste against my will." It is a pretty straightforward allusion to rape. It is also important to hear the rest of the girl's story: As he let her go (again, only at Zeus's command), Hades told Kore that she could do worse than to stay and be his bride. While she is there, she is royalty, Queen of the Underworld, and all would worship her. Note also that what he tricked her into doing was sweet, by her own description.

That part gets played down because the story is told from Demeter's point of view: its great version is the Homeric "Hymn to Demeter." Demeter has lost her child for a third of every year. That is why there is winter: for one third of the year, the pining goddess will not let anything grow. Kore, no longer a maiden, is now called Persephone, and when she returns to her mother—and to her family, the gods of Olympus— each year, it is springtime and the world rejoices. If it were called "Hymn to Persephone," it might begin when Kore first sees Hades approaching amid the fields of blossoms, luring her close with a false flower, then the terror, the abduction, the force or seduction, the sweetness of the pomegranate seed, the fear, the delight, the flattery, the pleasure in being far from home, being a queen, being unreachable by her mother. This is the story of a horrifying loss of innocence, but she wasn't only stolen and overpowered; she was also seduced and enormously empowered. All three names—Kore the virgin, Persephone the Queen of the Underworld, and Demeter the mother—were often understood as life stages of woman. As is well reflected in the biography of a Greek goddess, life for a woman of this world was about bondage to place and master, and it was about abduction, rape, and loss, or at least the threat of these things. There was a bittersweet motherhood and a brutal experience of familial and romantic love. Women's only source of control was the trouble they could make by freezing up in sorrow and refusing to bear fruit. Their icy silent treatments could make the earth cold. Now we know enough to go to one of their celebrations.

The Thesmophoria was an overnight outdoor party. Each polis had its own meeting place for this festival, and in various places it lasted from three to ten days. Athenian women headed to the Pnyx hill, near where the male assembly usually met.[5] It was within the city walls, but still remote. Sources suggest that only married women could participate—no children (except for a few nurslings) and no virgins. Husbands were required to allow their wives to go and were required to see that the festival was well funded. When the Thesmophoria appears in Greek literature, often some man tries to spy on the women and is either captured, killed, or castrated. Men wrote these works, and the women are support-ed for their vengeance. The Thesmophoria was important to the whole community; men, too, prayed that their babies would thrive, their wives be well, and their fields be fertile, and they believed that the women's secret rituals and bawdy frenzy could help. For the days of the festival most everything else stopped; courts were not in session, and all public business was closed.[6]

The first day of the festival was the "going up," when the women all climbed the great hill, with bountiful provisions, and set up camp. The second day was gloomy and hard: they fasted and impersonated Demeter mourning for Kore. They sat on the ground and wept. They searched the bushes and called her name. When it got dark, they lit torches and kept looking.

Then at last, later that night and into the next day (and in some places for days on end), the women feasted and cut loose. This part of the festi-val is called "the good birth." It was wild. The women ate breads baked in the shapes of penises and vaginas. They ate pomegranate seeds, which we saw were the image of semen in this myth, and they ate the roast pork that was the product of their earlier sacrifices. There was nothing very sacrificial about Greek sacrifices; they kept all the good meat for them-selves. It was a feast. When the women were not eating, they were acting out the later scenarios of Demeter's story. Ostensibly to copy Iambe (who cheered up sad Demeter while she was in the home of the king and queen of Eleusis), the women shouted sexually explicit jokes and insults, exposed themselves to one another, and spoke "obscenities." Some got naked and ran around. At the center of the celebration was an agricul-ture and human-fertility ritual involving the sacrifice of piglets, the scar-ing away of snakes, and the scattering of seeds. Everyone was responsible for bringing her own piglet. The pigs had several symbolic meanings, but were in part echoing a herd of pigs that fell screaming into the underworld when Hades abducted Kore. The ritual was full of arcane symbols, darkness, and secrets. The whole thing was an explicit plea, not for forgiveness, victory, riches, or salvation, but just for life. It was a ritu-al to ensure a year of good births, births of the body and births of the soil. When you have a life inside you, you hope it will be well, and you never forget the tenor of that hope. When you farm, your whole being hopes for fruit.

Today we worry about our children as much as ever. Think of those families where the daughter disappears and they search for her and hang "Missing" signs of her everywhere. After a while she is presumed dead, but she shows up, not having been kidnapped, but having been seduced. She wants to come home, wants to go back to junior high school. These families get their little girl back, but in a whole new universe. I believe that the reason our news is as interested in lost girls as it is in geopolitics is because the news is doing the work of myth for us, and when we respond with public grief rituals—vigils and memorials—we are doing what the Greek women did. Let's look at it from a more oblique angle. Imagine what would happen if once a year women met for a three-day reenactment of the drama of the lost child: maybe they take turns pretending to be Katie Couric while the "mother" tries to remain composed and say the words into the camera, *My girl is gone*, and hold up a photo. Don't ham up the scene, just try to get through the script, then go hang photocopied signs on telephone poles. Then have a vigil. Then a memorial.

Then the girl comes back. Her story is awful, but she seems okay. No one has hit her; they just pushed her around and seduced her and scared her. Your baby is back; not a baby anymore, but back. Let's say on the last night of this reenactment, the women partied. Each woman imagines a bed on the ground beside her, and into it she tucks in her imaginary girl, home again and safe. Each woman is then to drink and take whatever drugs she pleases, or none, and dance around that invisible bed, to go wild, scream with delight, fall on other women's arms. Even if no misfortunes ever came down on you, wouldn't this teach you something about how to mourn and maybe how to stop worrying? And wouldn't it bring you together with your neighbors? I think some people would look forward to this holiday all year long. I think some people would find it a relief to try on tragedy, and then have lived through it. Always bracing against pain is too exhausting; we need to embrace it now and again. It would bring people together in an intense and scalding way. It would let the mysterious power of the crowd develop intensely, to the point of euphoria, and it would let us have a public private life.

We know the reason that the ancient Greeks gave for their mad festival: It kept civilization going, it made barren things bear fruit. In our culture, some women who cannot conceive adopt a baby, and within a year they find themselves pregnant. The Greeks said that they celebrated because it helped their fertility. I'm sure it did. The wild celebrating of Demeter went on for over a thousand years and was the most long-lived ritual of the ancient world. Nothing lasts that long if it doesn't work. Enactment, doing, is more than you think. Thinking about being drunk, for instance, is different from getting drunk, and thinking about dancing until you are drenched is different from dancing until you are drenched. Letting go of your borders and boundaries and letting yourself be seen could relieve stress and increase fertility.

Imagine a mandatory, yearly festival where the focus is on babies. Such a festival could mean a lot to present-day men and women going through the experience of parenthood or its attempts. Even for women, our culture is blind to the strains of pregnancy, childbirth, motherhood, nursing, and weaning. The men who become fathers get yet less serious attention, though the experience hits them extremely hard, too. Want to know the leading cause of death of pregnant women in America today? Murder. The expectant dad kills her. That's a lot of tension. The ancient Greeks acted out this level of crisis for a short intense time instead of worrying a little bit every day.

Of course, all this attention to the loss of children was not just fear of parenthood. In ancient Greece, infant mortality was staggering. It is estimated that some 25 percent of all babies died before the age of one, and another 25 percent died by the age of ten.[7] Much evidence suggests that people loved their children then as intensely as they did in any other century. They just had a different range of realistic expectations for them. We all know about the infanticide of the ancient world. Sometimes families decided not to raise a child and instead put it out on a public hillside. But this was often not infanticide as much as putting the child up for adoption: many of these babies were taken home by someone else, often to be raised as a servant. In Greek literature, when people change their mind about a baby, they tend to go looking for her or him with the expectation of finding the baby somewhere among the population, and the babies tend to be found. The women's world was very marked by their babies, and very haunted with loss. But also haunted by recovery, in all senses of the word.

I am going to give you some details about two more festivals, and I want you to note whether any of the details, of the story or the ritual or the party, seem particularly attractive to you; whether you wish you could have taken part. Like the Thesmophoria, the festival of Dionysus was very ancient, and in its origins it was also for women only. Not one of the original Olympian gods, Dionysus, god of the bacchanal, was also called "Dionysus the twice-born." His worship came from outside Greece and caught on like crazy. One popular version of his origins says that when Zeus and Semele were expecting the baby Dionysus, Zeus killed Semele with a lightning bolt. Zeus found the boy's fetal heart and tucked it into his own thigh, where the baby healed and gestated, and whence he was born. In some versions, Dionysus's mother is Persephone (yes, Zeus's daughter), and after the child is born, Zeus's jealous wife, Hera, lures the boy away with toys and has him torn to shreds. His grieving grandmother, Demeter, finds the heart and nurtures Dionysus to rebirth. Lions and polar bears kill their cubs sometimes. Male mammals worry about how their stature will change as a result of a new generation. We are sophisticated animals, but we still sense the threat that we pose to each other, and we have the brains to worry it into the ground. For women this means a hostile world, but the myth also speaks to the women's own murderous frustration. Families draw on each other in complex, subterranean ways that no one ever fully fathoms. The Dionysus myth suggested them in vivid, violent terms.

The cult of Dionysus is about wine, sex, dancing, and madness. There was violence, too. Part of the festival was the *sparagmos*, which means "tearing apart," and the *omophagia*, which means "consuming raw": the women killed goats, fawns, and other animals, then tore them to pieces and ate the flesh raw. The women hunted for some of these animals themselves during the festival. Unlike the Thesmophoria, the Dionysian festival took place not inside the city walls, but outside them, in the wilderness. The women dressed in fawn pelts. At night they drank wine, sang, and chanted. We have vase paintings showing them dancing with their heads thrown back, throats exposed, eyes rolled back, and we have descriptions of them cawing, bleating, and roaring like wild animals. They feasted, and again they had bread and cake shaped like phalluses and pudenda. This time pomegranate seeds are forbidden. This is no fertility ritual: the women are in no mood to take in seed. In a rage, the father has killed us, and our baby. Then we are Demeter, the grandmother, finding our grandson's heart and healing him. The women acted out Dionysus's birth, death, and second birth. Dionysus is almost always surrounded by maenads—the word means "mad women"—frenzied female worshippers, and the celebrants took on that role, too. Later on in Greek history, Dionysus was surrounded by frenzied male worshippers as well. Greek theater originated in the performances that took place at the festival of Dionysus. The festival would also be a common setting for plays.

Euripides' *Bacchae*, the latest of the classical Greek tragedies, and one of the most psychologically interesting, is all about the Dionysian festival. A brash young king, Pentheus, looks forward to spying on the dancing women with the expectation of finding them "in the woods, going at it like rutting birds, clutching each other as they make sweet love."[8] In another musing, he imagines: "Mixing bowls in the middle of their meetings are filled with wine. They creep off one by one to lonely spots to have sex with men, claiming they're Maenads busy worshipping." But even this adolescent king, with his focus on sexuality, knows that, mostly, the women are engaged in wild dancing. The women were said to breast-feed wild animals, wolf pups especially.[9] We are in a world of deep woman-weirdness, self-exposure, and animal abandon. Here is Euripides' description:

> Some held young gazelles or wild wolf cubs
>   and fed them on their own white milk, the ones
>   who'd left behind at home a new-born child
>   whose breasts were still swollen full of milk.

It is nice to see the wolf show up this way, as a nursling of the mature woman rather than as the monster stalking the girl and grandmother. The women acted like wolves, in that they tore at and ate live flesh, and they made wolves human in nursing them. The image will return in reverse in the history of Rome, founded by the twins Romulus and Remus, who nursed at the teats of a wolf. It is a gesture that makes and unmakes civilization. The play also describes the festival's innocent euphoria:

> Then the bacchanalian woman
>   is filled with total joy—

like a foal in pasture
right beside her mother—
her swift feet skip in playful dance.

Euripides wrote that "after the running dance" Dionysus and his followers hunted goat's blood, "devouring its raw flesh with joy." Between the
running, the dancing, the sex, the drinking, the drums, and the violence, this is rowdy stuff. There is also much suggestion of the use of
powerful drugs.[10] At some festivals fermented grains were eaten to produce a hallucinogenic effect, we think due to a fungus that grew on the
grain. Euripides tells us that for the worshippers of "the god of joy," the
experience is delicious. For them, "The land flows with milk, the land
flows with wine, the land flows with honey from the bees." Milk, wine,
and honey abundance is here associated not with a future utopia, but as
a real description of how things feel when you get high and dance on a
hillside with your neighbors.

Euripides' play about the festival has a shockingly gruesome ending:
the mother of young king Pentheus is one of the leaders of the Dionysian revel, and when he sneaks into the worshippers' midst to spy on
them, his inebriated, indeed frenzied, mother tragically mistakes him
for a lion and kills him. It is only a play, not journalism, but the violence
is gruesome. In one of the real festival ceremonies, each adherent bathes
in the sea with her own piglet, which she later sacrifices. These are
almost all mothers, and they bring this little, pink, squirming beast into
the warm bath of the sea, and bathe it, and later kill it. It reveals an
incredible willingness to glimpse the dark emotions of nurturing love.
Later, there was another scene of mockery and obscenity that was said
to again be copying Iambe cheering up Demeter, as in the *Thesmophoria*. As the great twentieth-century scholar of Greek history Jane Ellen
Harrison put it, "Dionysus is as it were the male correlative of Kore, but
changed, transfigured by this new element of intoxication and orgy."[11]
These festivals of grief and triumph seem to have been a profound relief
in the short term and, in the long term, psychologically protective. At
enactments of the double birth of Dionysus, or in the abduction and
rape of Kore, mothers became embodiments of the sorrows of women
and goddesses. It was their job to act out the roles with passion and to
make the magic of rebirth come true.

How do we pray for fertility today? Consider another example: Beer bottles in America all proclaim that drinking alcohol during pregnancy can cause birth defects. That is true, and what it can cause is a sad, bad thing called fetal alcohol syndrome, where the child is born mentally disabled and facially recognizable as such. Of course we cannot do studies to find out exactly how much alcohol it takes and what other determining factors might exist (not every pregnant woman who drinks to excess will produce a child with fetal alcohol syndrome), so people figure it is better to be safe than sorry, and caution no alcohol at all. But I want you to consider how many people you know personally whose lives were harmed by their father's drinking while they were growing up, as compared to how many people you know whose lives were affected by anyone's mother's drinking alcohol during pregnancy. In the thousands of years of recorded history, in countries where they drink a lot, have we heard any association of the mother's drinking wine or beer or even scotch during pregnancy as correlated to the children's intelligence, success, and happiness? How about the father's drinking during a person's childhood? So why don't the labels say: "If you are a man and live in a family, it is advised that you refrain from alcohol consumption to prevent lifelong emotional scars on your children."?

This is about our society being much more comfortable limiting the liberty of women than it is limiting the liberty of men. But after all, it is just advice. Why do the women take it so seriously? They know that the Irish, the French, and the Russians have produced almost all the great novels, and they give their pregnant women beer, wine, and vodka, respectively. Many of today's mothers-to-be also know that their own mothers, and their husbands' mothers, drank throughout their pregnancies—Pink Squirrels and hot toddies just as often as they wanted. The whole thing is crazy. Alcohol is our culture's dominant legal drug. It has a lot of drawbacks. A mother who reports on a questionnaire that she resorts to our culture's big drug several times a day is also going to correlate with kids that have some difficulties, for other reasons. Scientists keep shocking themselves by finding ever smaller amounts of reported drink that correlate with kids having slightly lower IQs, but you have to think about what else you might be learning from a woman who is willing to say on a questionnaire that she drinks every day. I'm not telling you anything you don't know already. So why does the government put that warning on the bottles, and why do the women adhere to it far more rigorously than they adhere to any other rule about drinking or eating that they have ever been told? Many pregnant women wouldn't take a sip of champagne on New Year's Eve. Why? Because this is Demeter stuff. Let no distilled grain pass your lips for precisely nine months, and unto you will be born a healthy child. Amen.

The last Greek festival I promised you is not about fertility; it is about living forever. In the last stage of frenzied Greek parties, after the classical age of philosophy and the breakdown of democracy, finally, the guys get to be part of the bacchanal. After Alexander the Great, the Greek world got more cosmopolitan and the severe division between men's and women's spheres began to slacken. At this point in Greek history we suddenly see women included in paintings of fancy dinner scenes, and ideal love begins to be that between a husband and wife. Mystery-religion ceremonies—most famously, the Eleusinian mysteries—started as female but came to be a very heated coed escapade. Many husbands and wives were initiated together. Mystery religions were most popular with urban sophisticates. There was darkness and dancing, wine and drugs, esoteric rituals and liturgy. With the men involved, the key theme no longer had mostly to do with children, motherhood, and bountiful crops. Now it had to do with life after death; indeed, the mystery religions were a major source of modern belief in an afterlife. The other sources of this belief are Judaism, Platonism, and Buddhism. And of all these, the afterlife imagined by the mystery religions was the most personal: that is, you yourself were to survive death, as yourself, and it would be an almost physical existence full of insights and delights. I don't think the men caused the concentration on the afterlife, or that a switch of interest from birth to death made the mysteries appealing to men. Instead, I think something else caused both these effects: in the cosmopolitan empires of the end of the ancient world, men grew powerless. When the women were powerless alone, they reached for "other powers" alone, too, and they took the opportunity of being considered irrational to go off and indulge their wild desires. Now, as men were no longer citizens of the polis, but were now subjects of emperors, they, too, were powerless; they, too, were no longer the bearers of gravitas and decorum. Now they, too, could go throw their heads back and roll their eyes. Only a wealthy culture worries more about death than birth. Like the people of late antiquity, we today have a cosmopolitan assumption that life will get itself born and that food will be on the table. It is a good spot to be in. The presence of men at the Eleusinian mysteries, and the switch from the mysteries' being in honor of birth to being in honor of rebirth after death, makes it worth learning one or two details about this party.

The myth behind these nocturnal rituals was taken from an aspect of Demeter's story. As I mentioned, when the goddess Demeter was waiting for the return of Kore and was hiding out from the gods, she lived with the mortal king and queen of Eleusis, minding their baby boy. One night the queen was spying on Demeter and saw her putting the baby in the fire! Goddesses should not be judged as you might any other nanny: she was in the process of making the baby immortal. But the queen didn't know that, and she screamed bloody murder. Demeter was interrupted; and it ruined the trick, which couldn't be redone. Furious, the goddess told the king and queen of Eleusis they would have to build a shrine to her on their own lands, where she had searched and wept for her daughter. They would also carry out a series of secret gestures in the dark. This became the heart of the Eleusinian cult. The celebrations took place twice a year; the more important of the two went on for nine days and nights. The crowd acted as had Demeter, desperately searching for Kore, weeping and calling, there on the very ground where it all was supposed to have happened. But Dionysus was there, too, and he set the tone, with wine, theater, sexuality, and dancing. One of the most distinctive things about what went on was the secrecy. You were not supposed to reveal to anyone what happened at the initiation. That has made historians feel they do not know what went on there, but that is missing the forest for the trees. What went on there was: shared secrets. These secrets referred to the lost Kore and her return as Persephone, but focused most on what the goddess of grain, Demeter, was doing to the child of the king and queen of Eleusis. I can't help but wonder if the queen didn't have it right the first time. What a curse to saddle your son with! That he should outlive everyone over and over, and never get to rest. But most people do not feel as I do on this; they want Demeter to make them immortal. And these mystery initiates found the hook to hang their hopes on in the tiny part of the Demeter story that seemed to them to give evidence that if the goddess Demeter wanted to, she could make a mortal live forever. This was serious stuff. We have a letter from the writer Plutarch, who lived in this period, to his wife, written to console her because they had lost their beloved child. Plutarch says that their membership in a mystery religion ensured that they would all be together again someday.[12] What is crucial here is to note that whether they are dancing for birth or rebirth, these frenzied parties are dancing for something. Group wishing is not enough without group dancing, and group dancing is not enough without group wishing. The great happiness tonic is a large group of people who are both dedicated to a hope and willing to do some frenzied celebration and other festive ritual. The actual secrets in the Greek "mystery" religions were things like baskets being paraded around containing "no one but the high priest knows what." It was the sense of shared secrets, a willingness to go wild, and a dedication to a hope for salvation, for more life. This combination can be transporting.

How good was it? It reads as very satisfying to me. The event's great classic scholar, Walter Burkert, and the historian Jane Ellen Harrison both stated with confidence that the Eleusinian mysteries were occasions for cutting loose one's inhibitions to a profound degree. Burkert described the coldness of the sophisticated, individualist, rationalist life that was increasingly idealized in the classical age and explains that the Eleusinian mysteries were a response to constraining social mores. With the Eleusinian mysteries, "[a]n atavistic spring of vital energy breaks through the crust of refined urban culture. Man, humbled and intimidated by normal everyday life, can free himself in the orgies from all that is oppressive and develop his true self. Raving becomes divine revelation, a centre of meaning in the midst of a world that is increasingly profane and rational."[13] Jane Ellen Harrison wrote of an insider report on the Eleusinian experience: "The whole ecstatic mystic account beginning with the sensation of a blow on the head and the sense of the soul escaping, reads like a trance-experience or like the revelation experienced under an anesthetic."[14] Remember, it wasn't just a small sect of backwoods weirdos doing this; it was a large sect of hip, sophisticated weirdos. Refined, educated, financially comfortable couples dominated. The festivals went on for hundreds and hundreds of years, for thousands upon thousands of people. As Burkert wrote, the festival was an opportunity to break out of their rational, nonmagical world and go temporarily

crazy—to visit the mad side of one's mind. The mystery religions flourished throughout the Roman Empire, and wild dancing to Demeter had
started even earlier there. That's because in 496 B.C.E. there was a devas-
tating famine and the Romans decided to adopt the Greek goddess
Demeter as their own, calling her Ceres, the name of a minor goddess
of their pantheon. Ceres was the goddess of grain and of motherhood.
Hence the word *cereal*. At the festival of Ambarvalia at the end of May,
women celebrated Ceres, dressed as her, and carried out secret rituals in
her name. The Eleusinian mysteries ended in 396 C.E. when Goths and
other Christians destroyed the site and claimed the territory for Arian
Christianity.

Gatherings are a necessary condition for the survival of the group,
both really and magically. In ancient Greece, people believed that participation in the rituals described here kept the community going, magically. For us, if we join our communities' rituals, then our community
rituals continue to exist, with people like us in them. Community spirit
and team spirit are both called *spirit* for a reason. Community and team
are where we got the idea of spirit, and they are what still delivers. These
reenactment festivals—the Thesmophoria, the Dionysian bacchanal,
and the Eleusinian mysteries—are the events most often called *ecstatic*
in the whole history of the world. They helped cope with grief and
worry, and they were explicitly imagined as keeping the world going,
and keeping it fertile. Also, they provided the occasional euphoria that
people need to be happy.

18

Medieval Carnival

The medieval world and the ancient Greek world were both almost unfathomably foreign to our own and to each other. Yet both were our culture's ancestors. There are a few ways in which we are very much like the Middle Ages, and not at all like the ancient world, and these are manifest in their wild parties. What fueled the Greek women's rapturous ecstatic festivals were family terror stories, and true political powerlessness. Medieval festivals were, like ours, for both sexes, and for the general population (not just initiates), which makes a very big difference. Maybe you need a decisively limited group if you want to eat hallucinogenic fungi, engage in blood rites, and generally freak out all night. Still, if the medieval carnival was less frenzied and ecstatic, it was still freewheeling and euphoric. Today, our happiness parties have their own logic, but they get a lot less done than these older parties.

We have science and philosophy in common with the Greek world. What we have in common with the medieval world is Christianity, and a bit more equality for women than the Greeks had. More equality under monarch than under democracy? Yes. Though Athens was a democracy, it was a minority democracy of men whose families had been there for generations. That was truest toward the end, but from the beginning the polis always held more disenfranchised people than people who could vote.[1] In the Middle Ages, power was in the hands of a tiny top of the hierarchy, firmly at the peak because of bloodline, but with room for a lot of jockeying within that group: no one needed a résumé; you could get ahead just by being useful or amusing the right person. The queen's hair brusher—a princess, of course—could become her schedule maker, and thus one of the important people in government. Powerlessness on the bottom was equally general. There was no slave class in medieval Europe. Foreign families naturalized quickly. Men had legal dominion over women, but women could own things and land, run shops, chase their husbands with brooms, and tell dirty jokes in public. And since the average man had no political power, and most of what really went on at the top was about proximity and favor, men and women both had opportunities for resigning themselves to their fate, or for scheming against it.

There was tension between men and women, but it was mostly seen as domestic tension, not cosmic. In ancient Athens, men were citizens and thus expected to behave with decorum. It was for women to go mad, to become an animal and a violent spirit, to let blood run down their faces. In medieval Europe, partying was not about mad ecstasy. Instead, it was raucous, filthy, flirty, teasing, and soaked in alc. In this sense, we are more like them than like the ancient Greeks. There wasn't much of an option to not go to bacchanal or carnival. Carnival was part of the church calendar—skipping it would be like skipping Christmas. Yet it was profoundly mocking of religion. At the carnival Feast of the Ass, for instance, absurd "asinine Masses" were celebrated.[2] They were in remembrance of Mary's trip to Bethlehem—particularly, of the ass she rode in on. The day's fake priest would bray, and the congregation would bray back. As we look at what else they did, keep an eye on the relationship between what they said they were up to and what an uninformed observer would describe them as doing. Whatever else the carnival festivities were, almost everything about carnival was an acting-out of power inversions.

At carnival, there was permission to eat lots of meat, have sex, drink copiously, and laugh. There were topsy-turvy parades where everyone dressed and acted as their opposite, evenings of costumes, and Bible-reenactment theater. All sorts of folk games and delights were suddenly available, encouraged, and met by what the great historian Mikhail Bakhtin called "unbridled gluttony and drunken orgies."[3] In his words, people were allowed to dress as devils and run about freely in the streets and fields to "create a demonic and unbridled atmosphere."[4] Where did this fun fest come from? From the late 900s to the mid 1100s, Latin documents refer to an annual festival called *Carnelevare*—literally, "to remove the meat." It was a public feast designated as an occasion to settle debts, get them paid or forgiven. A nice idea. After all, it is arguable that paying back is more important than confessing; and that nothing is bigger than having the power to forgive. The feast took place around the beginning of Lent, and that likely accounts for the name *carnival*, because a main feature of Lent is not eating meat. At first they saw themselves as feasting in order to tidy up, to get rid of the forbidden item. Lent is a forty-day observance representing the time Jesus spent in the desert. It begins with Ash Wednesday and ends with Easter, when Jesus is said to have risen from the dead. The earliest mention of a parade through Rome in anticipation of Lent is in 1140, and we know they slaughtered animals for a feast. We're still not at anything like carnival, though. It is just a little holiday here. A few centuries later this becomes a crazed and general frenzy, sponsored by the church.

As early as the eleventh century, the little holiday came to be a particular site for religious reenactments, which became the birth of modern theater. What were the shows about? The key story of carnival was that of a young Jewish girl, Mary, whose belly swelled while she was yet a virgin. Her son Jesus became a preacher who said many beautiful and deep things. Among his followers, he had a special relationship with a woman with the same name as his mother, Mary. The preacher inspired so many people that he became a nuisance and was put to death. Three days later his friend Mary (Mary Magdalene) saw him walking around. After hearing of it, his male disciples had encounters with him as well. There was a sense, later hugely inflated by Paul of Tarsus, that the followers of Jesus would also live again after death. The stated purpose of the plays was to illuminate the facts of history so that everyone would know them and have an experiential link with them. Acting them out, and watching the plays, was a straightforward religious and social duty. But that is not all that was going on. Just as ancient Greek women acted out the stories of Demeter and Dionysus because they needed to act out powerlessness and hidden power, loss and recovery, medieval Europeans acted out the story of a boy without a father, who suffers terribly, and the pleasure and grief of his mother. Again, it is a family drama about powerlessness and hidden power, loss and recovery. These plays were presented in church, on Christmas and Easter, sticking close to the Bible text. They later were acted out elsewhere, on stages of their own, where they turned into dramas of contemporary danger and rescue. The entire dramatic activity of the French fourteenth century was devoted to "the miracles of Our Lady" in which Mother Mary appears and consoles or saves a suffering innocent—for example, a good girl who'd been married off to a lascivious aristocrat who was now treating her roughly. Occasionally, Mary saved a penitent sinner, but mostly it was a good girl, badly used by the world. Some forty-two of these scripts survive intact.[5]

In the first years of the 1400s, the first theater was set up outside the direct auspices of the Church in Paris, in an old hotel, and the fifteenth century became the century of the "mysteries." The actors were not traveling professionals; rather, each town had a club. The plays were very long; some took days. The cast might include hundreds of people, all earnestly playing out the roles of their savior's drama. There was a dedication to realism: when actors depicted the Passion (the torture and death of Jesus), they went through a lot of real pain, and there were many reports of actors playing Jesus on the cross nearly dying onstage. The audience, too, was unusually dedicated. For the annual reenactment of the Passion, almost everyone in Paris crowded into a vast theater to see it. Almost everyone. The city had to post soldiers to protect the deserted homes and stores against theft. The shows were intense, bloody, and by turns hilarious and bawdy. These plays were written mostly by priests, so they were entirely licit, but when one of your featured characters is *the devil* and another is *the fool*, you can keep everyone's attention with scenes of terror, pain, lust, sex, treachery, and perversion, and still be comfortably within bounds. Then there is the simple bloody sadism of the Passion. As we moderns have seen, the Passion of the Christ can still get everybody all excited.

In the 1400s the action at carnival was getting more topsy-turvy, more wild, more like the ancient bacchanalia, and something odd happened: people began speaking of the Christian *carnival* celebration as being a survival of the ancient women's festivals and the later mystery religions. It was not true. At least, it was not true in the way they thought it was; the carnivals that grew up in the Middle Ages were not a continuation of ancient rites. A breach of more than half a millennium separated them; for centuries upon centuries, one was already destroyed and the other had not come into being. Yet once this homegrown European holiday of abandon got under way, it became a new site for the year-round pagan rituals that had happily continued through the years, despite the prohibitions of the bishops. The only way the church ever got control of much ancient magic was by co-opting it into Christianity, and that is part of the energy of carnival. The rites that survived were, not surprisingly, about fertility: grain and babies. Demeter stuff. But, again, carnival was not a continuation of ancient rites; these just started to be carried out on Europe's festival of abandon.

By this time, theater included jokes about the central texts of the religion—the Nicene Creed, the Hail Mary, and the Our Father. There were parodies of the liturgy, and many have survived: the "Liturgy of the Drunkards," "Liturgy of the Gamblers," "Money Liturgy"; and parodies of last testaments: "The Pig's Will" and "The Will of the Ass." There was much tolerance for mixing comedy and fable with Biblical texts. People believed there had been a Pope Joan. As the story went, a smart girl in boy's clothing had worked her way up the Church hierarchy and was eventually elected pope, Pope John, all by her own merits. One day while traveling with her entourage, Pope John had to get down out of her carriage and give birth. Everyone was pretty shocked and angry. How angry? One version of the story is that they tied her legs with rope and dragged her around while stoning her to death. But in another version she went off to live in a convent and repent her sins, while her son grew up and became a bishop! She was never reverentially called Pope Joan, but got the name after she was defrocked. Today most historians say none of it ever happened, but it was a common motif of the period. A German play, *Frau Jutten* (1480), by a cleric named Theoderich Schernberg, tells of her as "an ambitious woman," and capable. Schernberg's Pope Joan submits to a rigorous penance and is ultimately saved through Mary's intercession. Acting out such fantastic tales, or at least seeing them, was expected of a good Christian, notwithstanding that they used the papacy for a melodrama and treated the mother of God as a soap-opera heroine.

That all changed with modernity. In the century beginning in 1500, a Roman Catholic split established Protestantism. With the arrival of

Protestantism, Catholic religious theater suddenly looked more like tomfoolery than religion. The Protestants criticized and rejected these fanciful plays, and then some Catholics started to worry about them, too. In 1548 the Parliament of Paris stepped in and banned the Passion plays. It was similar all over Europe. The plays stopped except carnival, where they expanded. As the new mood squeezed the rest of life, carnival got crazier. Carnival was not just a response to change. It was also a response to the repression that had followed in the wake of change. It was a hard time to live. You did not have to be terribly outspoken or extreme to find yourself in trouble and waiting to be tortured, burned alive, or beheaded. This time of change—let's say 1500 to 1700—saw the height of witch burning and the rise of a shockingly severe Inquisition. These barbarities took place in both Catholic and Protestant lands. Put in other words, the Middle Ages themselves were not as repressive and barbaric as was the transition from the Middle Ages to modernity (a period we call early modern). The same period hosted the most riotous carnival. The carnivals had mock courts and mock trials, mock torture and mock executions. In Spain they put pigs on trial and then put them through an "ordeal" and execution. An ordeal was the torture generally administered to people before they were killed for crimes—to make them confess their sins before facing judgment in the beyond. Modern readers may be shocked by the idea of women in the ancient world ripping animals apart at festivals; here in the regular life of the early-modern period, authorities literally ripped *people* apart, as a punishment. A mock authority merely reenacted it on animals at carnival.

In the condoned spirit of the holiday, carnival revelers chased their bosses and betters through the streets and whapped them with sticks. There was theater that might be pious or parody, but in a real sense there wasn't an audience. Everyone was part of the show. Goethe wrote that carnival "is not really a festival given for the people but one the people give themselves." Bakhtin wrote, "Carnival does not know footlights."[6] The crowd is its own spectacle. At carnival each year, a boy was elected Lord of Misrule. The laity provided an Abbey of Misrule; and among the printers' journeymen there were Lords of Misprint. Men dressed as women, women as men. In the carnival plays put on by the Abbey of Misrule there were all sorts of women behaving badly. In France, characters included Mère Folle and Mère Sotte: Crazy Mom and Drunken Mom.[7] Among French men there were appointed a Prince of Improvidence, Duke Kickass, Bishop Flat-Purse, and the Grand Patriarch of Syphilitics.[8] Jesters were proclaimed kings, and serving girls were queens. Even churches directly under the pope's jurisdiction chose a mock pope for the festival and let him satirize the pope without mercy. (Sometimes the popes did get angry.) Carnival meant inversion on every level. People wore their clothes inside out or wore blousy pants as shirts, pulled on over their heads.[9] They paraded in crazy order. Parading was a big deal in the Middle Ages and into the early modern. Today we see our towns and countries through statistics and news media. How could they see who they were? The answer is that they regularly organized themselves into a procession. The clergy of the town led, then the government, then the major land-and-money folk, then every guild, group, or association.[10] Everyone came out and had a good look at the town leaders, a good look at each other, and a few pints of ale. Processions are a recognition of official order. Carnival parade was the procession undone. It was a moving crowd, not a delineated hierarchy, and it evoked madness, fantasy, and inversion. People participated across significant economic differences, from noble to rich laborer to landless hired hands.[11]

Like the Greek festivals, carnival was about grief. In some ways it was the same old grief, but it was also new. Let us try to squint at the Christian story and see it in a new focus. Here is a family with two fathers: one is *the* father, father therefore of both the mother and the son; the other, Joseph, is the father of neither. Joseph doesn't actually show up in the Bible very much; just a few tiny mentions in Matthew and Luke, and even there, he is never even referred to after Jesus is twelve years old. Jesus is a boy with two fathers, and both are not around. As fathers they are not biological and not legitimate. That is the grief of the boy. Consider now the mother. Mary's coming up pregnant was a classic kind of *being in trouble*. In medieval Christendom, the most important images were the first and last moments of the life of Jesus. In both cases we experience the moment through Mary. In the first, a young woman is in labor in a barn. There was no room for her at the inn. People of earlier times did not need to work to remember the smells and the dirt in a barn, so it was a poignant tale. It would not be an easy birth for other reasons, too: not only was this her first child; the girl was a virgin. Her maidenhead was broken by her child, crowning. It is an image of an innocent and unprotected girl under extreme pressure. The last image of the life of Jesus was equally evocative. Here, the young mother is sitting on top of a hill where executions take place. Her son is already dead. They have taken him off the cross, and she is holding him in her lap, in her arms. The medieval mortality rate was bad for everyone. Mothers young and old held their dead sons' bodies in their arms about as often as sons, old and young, held the bodies of their dead mothers. The lives of Jesus and Mary were lives of familiar heartache, sprinkled with fantasies of acceptance, love, and triumph. Just like the reenactments of Demeter and Dionysus, reenactments of Jesus stories were a magnificent opportunity for emotion.

Carnival revelers indulged in gender-role reversals. We tend to see cross-dressing onstage or at Halloween as very different from doing it in your bedroom or at a bar, but it is arguably not.[12] Carnival was an opportunity to do what you wanted so it was a good time to discover what you wanted. In a Spanish play of 1611, young Serafina tells her maid Doña Juana:

At the Feasts of Carnival
Every woman ends up in disguise.
I desire to amuse myself;
It should not surprise you
That the clothing of a man attracts me
Since it is that which I cannot be.[13]

Carnival was rebellious; but it was also a sanctioned mock rebellion that served as a safety valve.[14] Two of the most famous theorists of carnival, Bakhtin, and the American historian Natalie Zemon Davis, saw it as much more an agent of change than a safety valve.[15] As Bakhtin put it: be they monk, priest, scholar, governor, or physician, at carnival, people willingly cast off their status in order to join the common melee "and perceive the world in its laughing aspect."[16] Zemon Davis demonstrated that the "women-on-top" trope of the carnivals, at the very least, reminded people that this "perversion" was an option. Charivari was a carnival-type event where costumed bands would go visit people who had made fools of themselves, particularly May-December newlyweds on their wedding night. They'd bang pots and pans under their windows—or some other infuriation—and refuse to leave until they were paid off. (There was often a standard price.) Charivaris mocked adulterers, excessive wife beaters, and husband beaters. The latter were called *skimmingtons* and were represented by men dressed as women, carrying the metal skimming ladles that symbolized a wife's weaponry.[17] Charivaris shamed the rebellious, but also celebrated them in a way, and advertised them.[18] Events of a carnival type, including feasts of fools, carnivals, and charivari, were a significant part of regular life. It is estimated that at the events' height of popularity, large cities devoted around three months a year to them.[19] Carnival mocked the whole order of things, mocked even the wrongness of sin and the rightness of virtue, mocked the pope with his permission and encouragement, mocked God in heaven and the devil in hell. One of the indispensable props of carnival was a wooden stage set of hell. Bakhtin wrote of carnival's hell theater: "The people play with terror and laugh at it; the awesome becomes a 'comic monster.'"[20] Like women-on-top, mocking hell is a complex behavior. It rivets attention on the threats of hell, but also elicits laughs.

Ancient festivals and medieval carnival have striking similarities. Both have two main sets: the world and the underworld. Both have girls impregnated by the uncle or the Father. In both there is the drama of the beloved son, his gruesome death, and his marvelous revival. Christmas is a time to meditate on his difficult but successful birth and the sudden rise in status of the new mother: kings now come to her and await her sign to see the heir to the world. Easter called minds to bloody death and magical revival. Carnival was a chance to try these on.

The age of witch hunts, Inquisitions, and religious wars did not last forever. After the transition from medieval to modern, the fear of change waned; the gross repression ended, and so did the festival frenzy. As always, while people are fighting about whether things should change or not, things change. Novelties become more familiar. When that happens, even the conservatives' lives bear no resemblance to those of their grandparents, and the energy of traditionalism falls away. In this case, in the 1700s, Voltaire and the rest threw a big public fit, telling the churches and governments that people aren't going to put up with this monomaniacal brutality anymore. As the Enlightenment took hold, carnival disappeared as a major force in European lives. It is too simple to say that when the tension relaxed, the wildness relaxed, for there were new tensions, of course, and new wildness. But a certain level of brutality was no longer condoned in the public eye, and celebratory bliss was becoming more private, too. Gone was the time of publicly turning the world inside out and upside down.

At the center of both Greek festival and medieval carnival were reenactments, and in both cases these reenactments were the birth of theater in their respective millennia. What an amazing thing. And when theater arises, general-population reenactments and costuming wane. Maybe as a civilization takes on age it gives up its exuberance and sets it onstage so we can rest our tired bodies and be entertained.

Today we watch an unprecedented number of people dressed up in costume and pretending to be other people. There used to be a lot of occasions for the general population to get into costumes; to take one more example, Jewish communities used to go topsy-turvy at the holiday Purim, where you dress up as the handful of members of the Purim story, and all the girls are the natural beauty and Hebrew lobbyist, Queen Esther. But nowadays we do not dress up in costume very much at all. What does it mean for us that we have replaced being in costume with watching people in costume? What happens to the populace when this shift takes place?

Well, what happened to the Greeks? As Greek civilization aged, theater developed and a certain kind of generalized revelry lessened, and as a response the mystery religions arose, creating new kinds of in-club revelry. The wild Middle Ages also saw general revelry cede its energies to theater, and by the end of the 1500s people were writing plays that we have not stopped talking about since. All this puts our television and movie watching in a different light. Screen games and Web surfing are perhaps halfway between watching and doing. As interesting as this makes all the entertainment, it is even more interesting to consider what we are missing by no longer actually putting on costumes. We are missing a lot. The power of costume is astounding.

We can glimpse this power by watching any makeover show on television. Someone is summoned because, according to their friends, they dress poorly. Often they clearly have issues: they don't like to be seen as sexual, or they walk around half naked and festooned with sexualized accessories. They wear very youthful clothes to telegraph to people that they are not part of the world of authority; they are unkempt to make sure no one depends too much upon them. We might guess that such a huge symptom would have to be addressed charily, through the emotions that created it, but, wonder of wonders, if you just force the person to put on more appropriate clothing, they look in the mirror and believe what they see. The twenty-eight-year-old who wore flip-flops to the office and carried a backpack as a bag cannot be verbally convinced that she is sabotaging her career, but the makeover people tear her down like an army recruit, and then put her in new clothes, cut her hair, get her cute shoes and a fabulous bag, hold her in front of a mirror, and compliment her until she starts twirling like a runway model. The whole therapy needn't take more than a day or two. It seems to stick, too, because it wasn't laziness or thriftiness; it was not being able to imagine themselves as a grown-up. Once they see it, they believe it. It can have an amazing consequence in people's lives. If you dress like a pirate you will feel a bit like a pirate. One indication of how much we lost with the loss of occasions for costume is how many of us want to be actors. It also helps to account for the oddly high status actors enjoy these days. They get our ya yas out.

Participation in Greek festival was mandatory; and carnival was an obligatory part of the church calendar. We moderns have to do this work for ourselves; the culture no longer insists that we show up to such parties, and it no longer provides such gala opportunities for costume. That is often how it is for people who live in cosmopolitan times: having eroded traditional rules and scattered traditional order, we get to live more freely. With our inner rubber band stretched so gently, we don't snap back into ecstatic (unless oppressed or a teenager). Today, for most of us, absurdity is the best we have, and most of it is contained in entertainment and art.

Get a good mask of whatever personally scares you and wear it out next Halloween. (Don't do anything to scare others while in costume, though. Virtue remains mandatory, even when you are masked!) How would it feel—good man or woman that I trust you are—to go around town in a mask, disguised as what scares you? It might feel very good indeed. What scares you is probably in you. Just as good as hiding what is usually shown is showing what is usually hidden. Imagine yourself in

a dimly lit ballroom with three hundred people who you know to varying degrees, and all of you are masked and disguised. There is music,
and most people are dancing. Some attractive body parts are on rare display. Most identifying features are hidden, but you know your friends
and acquaintances are there, somewhere. What would you be able to
enjoy about the experience that you would not be able to at an
unmasked party? In costume, the pressure to be you is lifted. We are
each a flutter of well-intentioned lies and postures, not least the face we
make when we are listening, and it is a relief to wear a mask. When the
costume is fierce, you feel a little fierce. Relieved and fierce is a nice
combination. If you are macho, it can be a pleasure to mince. The
licentious may find it scintillating to button up and act the virgin. If
everyone knows you as modest, it is gratifying to show them you know
about enticement. The rowdy young man may grow sedate when costumed as a Supreme Court justice. Costume offers an implicit challenge to the idea that people's daily roles are inherent to their nature. It
also offers a confirmation of these roles when, at the end of the festival,
having learned the weight of the robe, you are happy to cede it back to
the judge.

19

Today's News and Vigils

Similar to our political and social liberation, we have been liberated from nonsense. Again, our liberators left us in the lurch. Enlightenment thinkers proclaimed that rationalism and personal responsibility were better than obedience and wildness. It was almost immediately apparent to some that this would not do, and the artistic and social movement of Romanticism arose to celebrate the absurd and uncontrollable. For Romanticists, life was for passion and unpredictability, for tears and laughter, for believing crazy stories and submitting to love. Ours is a world keenly shaped by Enlightenment science, democracy, and decorum, but even in the world of the rational, scientific, and objective, we have found a way to tell ourselves the old stories.

Where, today, is the image of a crying woman searching for her abducted daughter, or the unexpectedly pregnant girl searching for a place to have her child? Where is the image of a sad young mother holding her son's lifeless body? Where are the images of relief when the sobbing mother is reunited with her child, who shows up shockingly alive? The answer is the news. We let the news anchors tell us stories of rape, murder, fire, insane mothers, cruel nannies, sly seducers, and violent fathers. This litany sounds different every year, but it is also very much the same, told in a style particular enough to constitute a genre, or a suite of myths. When you use a tiny bit of a huge amount of information (all events, ideas, or observations on Earth), the fact that that something is a *true story* matters, but we are responsible for the content in a way very similar to outright fiction.

Our version of mythos, the news, has some decided benefits, especially in that we are invited to be outraged and to do something about the information we receive. Part of what we learned from Voltaire was that the creation of public opinion depends upon information. It is up to each of us to be informed. Beyond demonstrating, or calling a senator, or writing letters to the editor, most informed people do not do much, and they may not do anything. They may not even vote. Yet the culture often reinforces the idea that the passive act of having an opinion is significant. We all influence one another, and the people who do get active on an issue may do so because they were emboldened by the quiet grumblings that they heard throughout their town.

Still, the news is such a huge cultural effort, and most of us do almost nothing with even the portion that can be called political. Why should the "political news"—economics, war, law—be partnered with deep coverage of the "popular news" of celebrity rape trials, an especially pleasing girl gone missing, a murderous husband on Christmas Eve? What makes them both news? They both happened, but so did a lot of things. Everything, really. So why are we assiduously keeping up on governments and pretty kidnapping victims? Because the news is our myth, helping us do our psychological work the way Demeter and Mother Mary did. Knowing the news holds a kind of cultural righteousness, yet it is not demonstrably good for much unless you get involved. Since most of us do not, it is reasonable to also wonder what we are really doing when we keep up on hard news. Part of the answer is that it makes us feel connected to each other. The soft news is even more useless in the ordinary sense of *use*.

Remember when the beautiful Mormon girl Elizabeth Smart was abducted from her bed? In her photo, she is posed with her harp. Her mother and father wept and searched. We had all seen girls disappear on the news, and after months go by, we begin to accept the worst. But then the Mormon girl reappeared! A reunion! But a lot had happened out there. Her mother and father are forever changed by their time of waiting and searching. The girl is returned, to be her mother's child again, but she comes back having experienced a version of adulthood. While she was still denying that she was Elizabeth to the police, she mentioned she had heard about that girl "who ran away." It suggests that this is how she thought of what had happened. At the time of her return, her father was quoted as saying, "I don't know what kind of hell she went through. I know very little about the last nine months. I have no doubt that she did fear for her life that night as she left the room, not knowing what would happen to her. She's grown a lot. She really is a young woman now."[1] Despite the fact that "been through hell" is a common phrase, this idea of the girl's time away is remarkably similar to Demeter's understanding of where Kore had been all that past winter. Patty Hearst, too, got stolen, but turned up seduced instead, having been a queen of the underworld. Then one day she's back, among with the wealthy celebrities again, those gods of Olympus.

Why are we so fascinated by the story of the lost child? The lost child is always you. Kore is always young Demeter, even if she is also Demeter's daughter. You are the one who had sex with Hades in the underworld and came home changed. Some part of you believes it wouldn't have happened if you hadn't been so beautiful and so easily deceived by fancy flowers. I am being dramatic, but what I mean is that losing and refinding is how you make progress in every human endeavor including the finding of oneself, and it is the drama of most love; progress comes in losing and refinding. It is also you who is running away, or hiding, in body or in mind. It is a risky game. Maybe no one will come after you. Though you work like mad to keep parts of you undiscovered, it is horrible to imagine that you will be completely successful. As the psychologist D. W. Winnicott wrote, "It is a joy to be hidden, but a disaster not to be found."[2] In a different context, Freud wrote that "[t]he finding of an object is in fact the refinding of it."[3] The way we know anything is at second hand; first we meet it and don't see it, then we come back and have a sense of knowing it. Some losses fascinate or appall because they happen so fast: one moment one world, one moment later the next. Most changes, however, are slow and awkward. Even fast changes usually take a long time: you lose your husband in a moment; you become a widow over a number of years. Life forces itself on you. Remember Kore, dragged "all unwilling" into a partial adulthood, becoming the queen of the underworld part of the year, and part of the year still her mother's daughter. The lost child is fascinating because she is you.

Why would an "educated person" in New York need to know about a particular murder trial in California? His pregnant wife disappeared. On Christmas. We hoped they'd find her. He searched for her, crying, but it was all a grisly charade. He had killed her and thrown her body—pregnant with his baby son—into the San Francisco Bay. Does this story sound interesting to you because it is a fact necessary to know in order to understand geopolitics? Or does it seem more like myth, like its function (in being chosen out of all the things that happen) is the same function myths have: to give us a context for our fear, hope, and anger? A trillion things happen every day, and the news could have developed in a very different way, highlighting very different things. But it didn't. The things it chooses to report, and the way it reports, have to do with what people are yearning to hear. Worrying over Elizabeth Smart, Laci Peterson, and Princess Diana is a public way to worry for oneself. These figures, along with the parents, the abductor, the husband, the royals, and the press, are characters for a psyche. They are mythic.

Princess Diana's husband didn't seem to love her; she didn't seem to have as much of her kids, or as much of herself, as she wanted. Her mother-in-law seemed cold and demanding. We all knew the photograph of Diana seeing her boys for the first time in a while and leaning down to scoop them in; and the one of her crying during a TV interview. She seemed to attempt to do charitable things. Then we watched her get divorced. We learned intimate weaknesses; she had tried suicide, practiced vomiting to stay attractive for the cameras. Then she was dead; it seemed the press had chased her into a wall, in a superfast luxury car. Dozens of the fellow human beings took pictures as she died. The last thing she saw was the strobe of camera flashes. Life must have flickered out like film from an old projector. We then learn of it by the news, like the family cat bringing in the family parakeet in its mouth, dead; the cat with a look on his face that suggests he knows it is a bad business. We had had a sense that time had been on her side, that if she waited it all out, eventually she'd be the mother of the king of England. Now it would never happen. In England it made them feel sad enough that they put down hundreds of thousands of bouquets around the palace. As the piles got bigger, some of us had a warm, hungry feeling, and wanted the flower piles to grow immense. The piles grew immense. The whole thing became an opportunity to commiserate with other people, and "show" the media how you feel, by tokens left and candles lit and milling around in the dark.

Not everybody liked it. Not everybody understood it. According to Martin Amis, in the *New Yorker,* no one did.

> No one understood it.... And we still don't understand it. My best guess is that the phenomenon was millennial. Human beings have always behaved strangely when the calendar zeros loom. And Dianamania bore several clear affinities to the excesses described in (for example) Norman Cohn's "The Pursuit of the Millennium": it involved mass emotion; it exalted a personage of low cultural level; it was self-flagellatory in tendency; and it was very close to violence. The phenomenon was, then, part of mankind's cyclical festival of irrationality.[4]

He could have given himself more credit; he understood a lot. It was a festival of irrationality. What many critics miss is that it was also the kind of "woman's story" that can call people's hidden grief out into public. Millennial zeros or not, Diana's death let out a blast of inner rage, empathy, and plain sadness that could seem "close to violence."

How did we get to the point where it seems like leaving a bunch of flowers is a big deal? Remember all the public drinking, sex, and dismemberment that used to go on at the bacchanalian festivals and the carnivals?! Not just in one or two towns, but everywhere, with everyone involved. Yet here, in response to floral pilgrimage, there were many scolds. Many people wanted to know why Princess Diana, who died five days before Mother Teresa, was so much more loudly mourned. People show their mutual grief because they have mutual grief; they show it in these eruptions when there are insufficient ways to show it scheduled into the regular calendar.

Consider this nasty mock eulogy: "The Princess ... is dead. She no longer moves, nor thinks, nor feels. She is as inanimate as the clay with which she is about to mingle. It is a dreadful thing to know that she is a putrid corpse, who but a few days since was full of life and hope."[5] This time it is not Diana. The princess referred to here was Charlotte, and our curmudgeon is the English poet Percy Bysshe Shelley. The year was 1817. Shelley was angry because within a day of the princess dying, three men—Brandreth, Turner, and Ludlam—were executed by the state, and Shelley thought they were more worthy of the country's tears, which all seemed to be going to the princess.[6] The men's crime was called treason, but it appears to have been a protest against local job conditions: people were working fourteen-hour days and not making enough to feed their families. There were reputedly four leaders, but the one who instigated and orchestrated the whole thing, Oliver, turned out to be a government plant. Here, Shelley said, was a true tragedy. Many agreed, and many also knew that Princess Charlotte was not being mourned for being a person of exceptional character. Rather, as a princess, she was known for almost entirely domestic aspects of her life. Daughter of the Prince of Wales and heir to the throne of England, she had won much public attention when she refused her father's first choice of husband for her—a prince famous for his drinking—and married for love a few years later in 1816. In the first nine months of her marriage she had two miscarriages, and then, late in 1817, she gave birth to a stillborn son and died herself the next day. She was twenty-one.[7] The philosopher Harriet Martineau witnessed what followed in the country and described the death of Princess Charlotte as "the great historical event of 1817." Moreover, "never was a whole nation plunged in such deep and universal grief. From the highest to the lowest, this death was felt as a calamity that demanded the intense sorrow of domestic misfortune."[8]

Shelley claimed that he couldn't understand why people would so mourn the princess. After all, he said, women die in childbirth all the time. These were nameless strangers to us, yes, but who among the mourners had ever really met the princess? Shelley reminded his reader that every day some family loses its wife and mother, often in poverty. "Are they not human flesh and blood? Yet none weep for them—none mourn for them—none when their coffins are carried to the grave (if indeed the parish furnishes a coffin for all) turn aside and moralize upon the sadness they have left behind."[9] Shelley did not see that in mourning the princess, the people were mourning all their dead young mothers, and expressing their own fears. People need public expressions of grief and fear in order to be happy; this kind of psychological work cannot be done alone. Vigils and memorials can be cliché, can feel shallow or scripted, but there is no need to question the sincerity of the grief. It is appalling when people hold up public vigils to a kind of grief detector, as if the crowds might be playing at this, as if there could be a "playing at this."

There have been crabs who were against the Dionysian bacchanal, grumps against the carnival, and grinches to curse even our sedate modern holidays. But life is reinforced through these collective experiences. We have to get our fill of them, and we have to keep an eye on them. We should also keep an eye on who the crab is. Sometimes he or she is blessedly free of the burdens that are relieved at these experiences; sometimes he or she is alienated from the experience by gender, class, or personal taste. Sometimes, of course, he or she has a point, and should be heeded. Still, there is an odd amount of anger against this emotional mummery. What Shelley and many modern critics miss is that the great number of women who died in childbirth, often leaving orphans, contained a sadness apart from politics. Brandreth, Turner, and Ludlam, and Mother Teresa, were hailed by many as symbols of liberty and of selfless generosity. But these figures did not strike personal chords among the mass of people in the way the princesses did. Most of us have not tried armed revolt; most of us do not serve the poorest of the poor; but everyone has worried over someone's bad husband, someone's childbirth, and someone's death. Charlotte and Diana were potent symbols of family grief, and there was both a sting and a comfort in knowing that calamity happened even to beautiful princesses.

There were a lot of grand, extreme funereal processions in the Victorian age, including one for Victoria herself, at the end of it. Victor Hugo's funeral shocked everyone with its excesses as people flowed through "all the streets of Paris" following his funeral train. Apparently more people showed up to mourn him than the number that usually resided in Paris.[10] There were poems and flowers and thousands of embroidered remembrances. How we mourn is a big deal. For most of the nineteenth century in Europe, public grief strove for excess. As the twentieth century opened, people began to reject many aspects of Victorian treacle and to insist on keeping funeral rites simple and noble. In the United States, the immense death toll of the Civil War brought a new dignity to burials. Our somber Memorial Day began in 1865, to help handle the unspeakable loss of life. There were just too many to be flowery about it. Now the graves of our soldiers are simple crosses, stars, and crescents. In Europe the change did not happen until they, too, were faced with massive death tolls, in World War I. In both cases, it began to seem gross to respond to the death of a celebrity with Victorian abundance when so many ordinary men and women were dying violent deaths every day.

In Europe and America, for most of the twentieth century, public grief over famous deaths was restrained, and the funerals were dignified. The most prominent death of what I think of as the short twentieth century (1914–1989) was that of President John Kennedy. His youth added to the tragedy that people experienced—and to the sadness that some people feel about it today, sometimes without knowing much about what he represented and what kind of times those were. His funeral was emotional, but its emotions were contained in somber ritual: the riderless horse, the salute given by his young son. It was poignant but austere. Not until the 1990s, again starting with a dead princess, did ornate public weeping, scribbled notes tucked into fences, and otherwise flowery grief come back in. The twentieth century—shocked by the scale of modern death—showed some gravity in funerals, memorials, and other responses to the news. The news itself was manly, and it generally called for a manly response. The young twenty-first century, like the nineteenth century, allows itself more feeling.

This rise in public mourning since the death of Princess Diana is a much discussed phenomenon in many countries, especially in the English-speaking world. An article in the Australian newspaper *The Age* discusses the new phenomenon of effusive mass public mourning.[11] It cites these interesting rituals: joining hands in a great ring of strangers; "paddling out," a surfing ritual derived from a traditional Hawaiian mourning ritual; vast moments of silence; and, finally, one in which you are "invited to wear a piece of wattle as a tribute to the dead."[12] (Australia's "floral emblem" is golden wattle, whose clutches of puffy flowers look like lemon-colored grapes.) In response to the tsunami of late 2004, another Australian newspaper tells us that the country's biggest ecumenical service is to be held because these events "spoke across the divide" of various sects. A preacher explained: "Something like this transcends religion—grief is an experience we journey through together."[13] We are alone in so many ways; we do not worship in temples together, we neighbors, we countrymen. But we do watch the same sitcoms and the same news, and through them we sometimes suffer together.

Journalists in various countries note the rise of public mourning and list the events that have recently brought on an orgy of mourning. It is heartbreaking how obscure some of the local dramas are from the perspective of the other side of the world. Everyone mentions Diana and 9/11 as the starting points, but in England they also speak of the astonishing public reaction to the murder of schoolgirl Milly Dowler. In Ireland they also speak of the outpouring of grief for the journalist Veronica Guerin. Australia is still marveling at how, in January 2004, everyone broke down when the cricketer David Hookes, apparently attempting to quell a bar brawl, was beaten to death by a bouncer. During the aftermath of the tsunami, in December of the same year, one Australian journalist wrote an article called "What Made Us Cry Such Bitter Tears for Hookesy?" about the strange level of grief displayed a year earlier when Hookes died. People all over Australia had worn black armbands and set up shrines. In recent years England has had a few massive scenes of the communal outpouring of grief, particularly grief over murdered schoolgirls. Along with Dowler, there was "the Soham tragedy" of 2002. Two little girls disappeared; they were best friends and in pigtails, and the news showed a photograph of them together, wearing David Beckham soccer shirts. They were found too late, having been raped and murdered. People traveled across England to Soham, where the girls had lived, to put down flowers and cry together, sign condolence books, and leave notes of other kinds. There were crowds of candlelight in the dark. Beckham's team, Manchester United, is as well known as the New York Yankees, but people who thought of the footballers as their team still mention that fact as a personal connection to the girls, whenever these murders are mentioned. It felt personal.

I am a New Yorker, and I can attest that if, just after 9/11, there had been any ritual associated with keeping ourselves and our city safe, we would have gone and done it. What we did had its own truth. We walked around town and dutifully read the wrenchingly useless "Missing" signs. We wrote notes on paper memorials in Washington Square, Union Square, and Central Park. We walked from vigil to vigil, stunned and bearing witness. That is how a solitary person like myself ends up in the park holding a candle and crying, hoping these strangers around me, whom I now sort of love, won't spill too much hot wax on my coat and shoes. We read the profiles of the lost lives in the *New York Times*.[14] An untold story is that all that year (yes, we felt markedly strange for a year), it was a relief to have a reason to be sad together, rattled together.

As much as we should allow ourselves to connect to such stories and to the crowds they draw, we should also learn to see them for what they are: true, as fears go, but not likely to come true in your own biography. When we watch the news, we should remember that most little girls are not abducted or killed. We are being told about the ones who are abducted and killed because we are scared of it, and that makes us interested in it. We need to take that experience seriously, but also to remember that the news itself is a little mythic, a little false, edited down to what catches our attention. We need to notice which stories catch our attention. We need to let ourselves feel things, but not be unaware of how symbolic the whole thing is. Imagine a button you push and the world for a mile around you goes dark except for anything that could pose a safety threat, and the threats glow green. Some people would almost never push the button. Some people would push it a lot. Some people would tape it down and live in a world made only of darkness and threats. That person thinks herself or himself cautious, but has made a choice to live in a terrible world. Why would anyone do that? Because while you are actively evading dangers, you don't have anxiety. Anxiety is what you feel when you are trying to be normal on the outside while your emotions roil within. Once we are screaming at the top of our lungs, we are no longer anxious. Likewise a gaggle of worries is a relief in comparison to fear of "whatever it is that is coming next," which is what we should be afraid of, and about which we can do nothing.

It is hard to know how to feel about ourselves as part of the news audience and, occasionally, as the object of its attenton. It is not all bad and it is not all good. The worst part of it is that it uses real people, who have to bear becoming public curiosities and caricatures, at a time in their lives that is often surreal in its misery. The best part of the way our news works is that sometimes people respond to what they learn and make the world better. Apart from the worst and the best, there is the way, for better or worse, the news gets used by most people, most of the time. We broadcast sad stories, and some reunions; death and a few revivals. When people need to meet in the streets over it, something in that is central to political identity, spirituality, and to theater. It is how we discover that we have a common heart.

20

Weddings, Sports, Pop Culture, and Parades

There are important similarities between the roles of Demeter, and Mother Mary, and the TV news's "mother of the missing child." People need community events—for both psychic and political well-being. I want here to stress that sometimes the culture's central generative story is connected to a community's big gatherings. In today's culture, story and gatherings are not often in sync. We are worried about bombs and disease, but at holidays like Thanksgiving or Labor Day there is no room for the expression of dark emotions. Even Memorial Day, despite the name, is all hot dogs and hamburgers to most of us. Certainly the public mourning of someone we all knew (and none of us knew personally) is one of the most direct ways that we connect our story to community events. The ritual of mourning for famous people is a rare moment when our most powerful stories intersect with a sizable gathering, and it comes to hold a lot of communal emotions.

Mourning rituals are not the only way we moderns come together and share. In our culture, most of us expect to be the center of one gala party that may cost about a year's rent: the wedding. Gala weddings are a performance of abundance: a dream world of food, sometimes displayed as if to make us think of Eden, heaven, Israel, and the Land of Cockaigne. It is not just that the food doesn't run out; it is that the display is almost transporting: the wealth of carved meats, exotic stations for vodka and caviar, food sculptures, chocolate fountains, and mountains of pastries. Note that it is generally only the first wedding that deals in this kind of fantasy. Think about the food abundance here, what it means to the eye and mouth, and the promises that are being made about love and happiness. The cornucopian food helps us suspend our disbelief, creates magic that outshines statistics. Second marriages cannot claim the same perfection (given our ideas about what marriage is), so they don't; and since they don't, we don't need all the food. But, of course, you need the cake; even a fifth marriage will usually get a tiered cake. Happiness needs cake. For the principals, there are rituals and costumes. For everyone, there is access to a special kind of dancing. One part of it is circle dancing. Think of Matisse's *Dance*, which pictures five men and women holding hands and dancing in a circle; the joy is transporting.[1] As the novelist Milan Kundera wrote, "Circle dancing is magic. It speaks to us through the millennia from the depths of human memory."[2] As a happiness lesson, nothing could be more straightforward: if you get a chance to dance in a circle, get up out of your chair and do it. Walk over and put out your hands, and people will unlatch themselves to take you in. You cannot just think about it and get the benefits, as you cannot just think about whiskey and get drunk. I don't want to hold two random people's hands either, but the simple advice of the ages is that when you get a chance to dance in a circle with people, get in on it.

Also important is joining in when people start acting out the letters "YMCA." I am joking a little here, but the more impossibly stupid it feels to you, the more I insist that you try it. It is great for you if everyone is doing the same motions simultaneously, but the fast disco wiggle interchanged with the slow couple shuffle is as much a cultural creation as any other ritual dancing. This dance act seems like a default remainder of skills that were truly practiced in the past, but that view may be exaggerated by our perspective. A space alien pointing her telescope into the party window would see a group of people rise from their chairs, gather in the room's center, and become an unpredictable but controlled machine of movement, rarely colliding. If our alien studied it, she'd see that the dance was governed by rules: who gets touched, where and how much, what facial expressions are appropriate, and, for instance, what kind of dance moves could get everyone else to stop dancing and watch. Our alien need not be from outer space. An untrained Frenchman or a Brit dropped into an American wedding often does not know how to negotiate it and stands about watching. It is worth noticing that as ordinary as they are to us, our celebrations are complex affairs for which we have a lot of privileged information. It is good for us to notice that there is a dance to be danced and that we know how to do it. Also, it is good for you to be in a crowd where most everyone has taken the same drug; to do this legally in the United States, the open bar at a wedding is your best bet.

For a demonstration of all these marvels, see the opening scenes of *The Wedding Crashers* (2005), in which two extremely charming and appealing youngish men (Vince Vaughn and Owen Wilson) use wedding season as a chance to meet women, sleep with them, and never call. The movie's tour of American weddings is brilliant fun, because instead of seducing only individual women, the guys seduce the whole party, and the women shake loose like apples. The way they seduce the whole party is by eagerly participating in all the rituals and thereby making the rituals seem youngish, charming, and appealing, which gives other people the permission (or even the idea) to gleefully embrace the round of rituals themselves. The plot of the film, unfortunately, is a tale of how romantic love stops the guys from doing the very thing the film had to offer. The plot should have been that romantic love, once attained, allows them to realize that what made the weddings fun for them wasn't only the aroused women. Coupled up, they could continue their spree. If you went to a friend's loft party, with a hundred strangers, you would not leave your purse on your chair and go dance. Yet at a wedding where you know only a handful of people, you very well might. It has nothing to do with knowing the people. It is that the culture of the wedding includes a letting down of one's guard, such that stealing would be experienced as especially low by the stealer. Because it feels safer, it is safer. Viewing a wedding with an anthropological eye, with historical perspective, cannot pull you *out* of the event if, prior to taking this view, you were held back from participating at all. People who sit out our tribal wedding rituals do so in part because they have no sense of these acts as tribal rituals. If you were in some remote land and the people invited you to drink some wine and do their dances in order to effectuate the social magic of turning two young people into a single adult couple, you would go, and join in, and feel elated. Nothing stops us from a similar elation right here at home.

In the section on money, we saw that music culture, sports media, and television shows provide, for many of us, a replacement for the missing middle of society: they give us shared experience and something to talk about. In this chapter we are interested in festival euphoria rather than daily public life, so here shopping is less significant. Our quarry here is the rarer, more emotive occasion. My criteria for a festival experience is that it be a large gathering with most of the following: costumes, ritual gestures, music, reference to sexuality, expressions of loss, expressions of victory, at least some hollering, and unusual touching. It is good if the participants come from the same place, such that they might see their neighbors, doctor, clients, and friends. It is best if it is a public event, cost free and open to all. It is also best if it refers to a very visible, acted-out myth or drama (as does a wedding), or if it draws on current events. It must also be safe for girls, which, given all the dramatic debauchery, requires that the festival itself be respected, as if sacred. Nothing, sadly, entirely fits the bill, though many things come close, but I have chosen a few gatherings to analyze for their aspects of festival euphoria.

Attendance at a sporting event is an occasion close to festival. You get to show up someplace with a lot of other people and emote. There are underdogs and heroes, jinxes to be overcome, personal feuds, and bad boys. The meetings are always the same, but different. Spectator sports work even better than religion in some ways, because they have the dopamine perks of gambling. Your team may win. You can go alone, or with your family, or friends. If you have season tickets, you get to know the people in the seats around you. If it is football, there are tailgate parties. Many people have rituals that they do to make their team win, so they feel actively involved. Every game is interactive theater, with special clothes to wear, and, for some, colors to paint your face. The "wave" is a remarkable social gesture where the arena sections take turns standing and lifting their arms, so that it looks like a current of energy flows around the stadium. The wave happens several times at almost all games, randomly, when a few fans get an urge to start one. We saw that being a fan creates the sensation of belonging to a vast brotherhood. The sports experience is a moment for drinking alcohol in the company of this brotherhood; suffering with others when your team loses and, when things go amazingly well, celebrating with hand slaps, and even hugs and chest butting. You get to punch the air and scream. There is the seventh-inning stretch, and the sing-along of "The Star-Spangled Banner" and "Take Me Out to the Ball Game." Of course, showing up is fatiguing, there is traffic to consider, not to mention loud fans, overpriced food, and a fixed view of the game. But there are good reasons to show up anyway. It is not for everyone, of course, but it is useful to note what these events mean—how they work in the same ways as historic festivals, occasioning expressions of sorrow and triumph.

Music shows have festival characteristics. In the same years that the American nightly news was developing an ideal of intense seriousness and masculine objectivity, *The Ed Sullivan Show* gave us the Beatles surrounded by American girls screaming their heads off in an agony of euphoria. Rock shows took on the attributes of ancient and medieval community festivals. The music is emotional and dramatic. People take drugs. There are outfits to wear. There is darkness and spooky lights. There is often reference to death. Performers and spectators alike wear religious signs. There is opportunity for primal scream, trance, or sexual interaction. There is imagery of devils and of flames. Individual fans hold up a flame.[3] As in the ancient world, it is mostly the women who scream, but anyone may dance alone or with others, as wild as they want. At goth shows, anyone may arrive wearing makeup and in costume; at all shows people wear garments naming favorite performers. Many concertgoers will sit still and watch the show, but many will move a bit, yell song titles, and show enthusiasm when they hear a song they recognize. Opera also has mythic stories, disguise, darkness, violence, sex, and costumes, but most of it is onstage.

Sometimes television fandom reaches levels of festival experience. The TV show *Star Trek* ran from 1966 to 1969. It was a fantasy adventure about space exploration, set in the future, but marked by the "sixties" idealism of its moment, and the secular humanism of its founder, Gene Roddenberry. Its fans, often called Trekkies, gather at conventions, where they dress up in character costumes from all over the universe. Denise Crosby's Lieutenant Tasha Yar was one of the original crew of *Star Trek: The Next Generation*, the first remake of the classic show. Crosby left before the end of the first season, but she made an impression on viewers. Yar was a well-developed character: a women in a traditionally male job, head of security, she was short-haired and curt.

She had gotten that way because she had grown up in a devastated place, scavenging for food and dodging rape gangs. Also, in her few episodes, Yar had sex with Data, an android. Soon after, Crosby's character was killed off. It is a much more mythic and memorable backstory and narrative arc than most of the show's characters have, and fans grew attached to Yar, especially after she was dead. When Crosby first appeared at a *Star Trek* convention, the admiring response surprised her. She had not realized that she (the actress and the character) had become a legendary figure in a strange midsize polis.

Crosby has since produced and hosted two terrific documentaries on the subject, *Trekkies* and *Trekkies II*. They offer keen evidence of how people negotiate our fractured, overpopulated, commercialized culture. *Star Trek* conventions create something like an Old World village, thick with midsize associations and personal interactions. There are huge merchandise rooms where people exchange Trekkie products and banter about them. Strolling the convention hall, they see friends, acquaintances, and group heroes such as "Commander Bobbie," who made the national news for wearing her Star Trek uniform while she sat on a real-world courthouse jury. There are memorabilia auctions, where you get to show people what you like. Publicly handing over fourteen hundred dollars for a Klingon forehead prosthetic worn on the show vividly enacts the group's values. It is roundly applauded. The highlights of the *Star Trek* conventions, of course, are the lectures and interactions with the show's stars. Many of them have individual fan bases. There are Spinerfems: women who adore the actor Brent Spiner, who played Data (a character who is sexual, but explicitly *not a man*), a very nonthreatening imaginary boyfriend to choose. Then the fans publish their own scripts and stories, especially now, on the Web. There are thousands of such fanzines. In them, there is much reference to Klingon fighting sex, and also much reference to the sex between Data and Yar. The show did not harbor much genuine darkness. What darkness exists is mostly cosmopolitan pain: Spock's Vulcan/human origins, Deanna Troi's Betazoid/human origins, and Worf's Klingon biology and human upbringing. There is no shadowy worry of meaninglessness, and little real angst over our animal constraints, pain, hunger, and dying. There is no wolf. Worf comes closest. He and his Klingon ilk are a favorite costume. Again, this behavior is not going to appeal to everybody, but apparently there is a *Star Trek* convention every day, somewhere in the world.

Cast a quick glance at the older costumed tradition that was *Rocky Horror*. At first in Greenwich Village, and later all over the world, people would repeatedly go see the film *The Rocky Horror Picture Show* week after week, usually at midnight on Saturday night. There were scenes based on cannibalism, sex of several varieties, and general cinematic darkness. The reason this was compelling week after week was that many audience members dressed in the outrageous underworld outfits of the film's characters, and a few acted out scenes, up front below the screen. Many brought bags of props, so that throughout the song "There's a Light," the audience twirled flashlight beams on the ceiling. For the wedding scene, they tossed rice. Since the audience knew the lines of the film, they inserted dialog so that the characters seemed to be conversing with the audience. This dialog was always developing: you shouted out your idea, and maybe it got a laugh and maybe it would be repeated, by someone else, next week. A proud moment, silly but satisfying.

For a bigger party that fits our criteria, consider New York City's Greenwich Village Halloween parade. What is always great about New York—the random brilliant jokes that its citizens tell each other in the form of art, gesture, and dress—is in top form here. The parade is a reckoning, because people dress as components of the major events of the foregoing year. (You can see thousands of pictures of the parade on the Internet.) When our son was five months old we dressed my tall husband as the Empire State Building, me as Fay Wray, and the baby as King Kong. Wray had died that year. We were a big hit. My husband carried the boy monkey on his shoulders, and would raise the baby's little arm in mock anger, and I would mime movie-poster fright. Decades ago, when I marched in the Village Halloween parade for the first time, my friend Mary and I just painted our faces to look ghoulish and wore black. If you want to, you can stand out and blend in at the same time, as there are amazing bands playing and marching, and if you march with them, you dance with their crowd. It is nice to be in disguise, among other people in disguise, with everyone dancing, invisible in plain sight, making the revel with your enthusiasm. Think also of Lou Reed's song "Halloween Parade" of 1990, in which the parade has become a time to notice which friends and beloved local characters had been lost to AIDS. An annual parade can be a yearly weighing up, show-ing who we are and who we are missing. Overall, though, the holiday is not somber; it is joyous, full of absurdity, inversion, topical commentary, and topsy-turvy nonsense.

Those who attend such an event, blocked off by temporary railings and lots of cops, are not in the parade, but they are not just spectators, either. This is no stage: the audience is half the experience. They are piled high on the sidewalks; they are sitting on mailboxes; they are in every window, on every balcony, and on every rooftop. They are every-where, and their cameras are flashing, and they laugh when you act out your character, and you laugh, too. Most secrets to happiness take ongo-ing dedication. But public celebration need not be done very often to get the benefits. Find a festival for which you have some affinity, and go. Don't drink so much that you numb yourself to the emotional experi-ence of being in the crowd. If you can, try also to attend a festival as part of the show rather than as a spectator: wear a costume, sing along, dance. Go to any crowd where you agree with the reason for crowding—be it politics, music, sports, or holiday—and become your own experi-mental animal. Consider your emotions at such gatherings, in terms of both moral feeling and mood. Think about the things that the Eleusini-an mysteries gave participants, and what carnival seems to have done for its celebrants. Are you getting these things? Do you want to? Have you been euphoric in public lately?

I think that we need this, we do not get enough of it, and what we get is in the form of absurdity in art. Complaining about culture having ran-dom, intense rules of conduct is like complaining about the inconve-niences of gravity. All we can do is try to find a way to alleviate the pressure now and again. That is what topsy-turvy does. Our arts reflect our hunger for it. Often they do a great job of providing it for us, singly and in communion with others: we share our love for the topsy-turvy by citing our love for artistic expressions of it. Think of the special kind of allegiance people have to films that are absurd: *Monty Python and the Holy Grail*, for example, or *Catch-22*, *Slaughterhouse-Five*, *Pink Fla-mingos*, *Being John Malkovitch*, or *Donnie Darko*. We need play, and the play has to be daring; it need not scatter all meaning, but it does have to turn things upside-down.

We get our wild festival and carnival topsy-turvy at the movies. *Willy*

Wonka and the Chocolate Factory is perhaps the greatest film of food abundance, a brilliant enactment of the Land of Cockaigne for the industrial age: the delicious factory. Wonka presents his guests with a river of chocolate running through an edible Garden of Eden. There are also "everlasting" candies, a stick of gum that tastes like a multi-course meal, and soda pop that can make its drinker fly. Along with showing us all the abundance, the film is something of a scold, which is a combination we are used to from carnival. The unusual and unexplained abound, from the Oompa Loompas to Wonka's office where everything is only half of itself—half a clock, half a desk, half a coatrack. The tropes of topsy-turvy and food paradise are kept in tension by the sour charm of Wonka. The boat trip he takes his guests on is menacing: they enter a dark tunnel, accelerate roughly, and Wonka sings:

> Not a speck of light is showing,
> So the danger must be growing.
> Are the fires of Hell a-glowing?
> Is the grisly reaper mowing?

A mean song sung sweet, perfect for a carnival night. Wonka's acidity demonstrates that he is not a foolish man being careless, but rather a serious man being absurd. As he puts it, "A little nonsense now and then is relished by the wisest men." The film's fairy-tale ending is that Charlie gets the factory and is to bring his whole family to live there.

> Willy Wonka: Don't forget what happened to the man who suddenly got everything he always wanted.
> Charlie: What happened?
> Willy Wonka: He lived happily ever after.

We know how you end up living happily ever after: you start from too little to eat. But no matter what they start from, most people like to daydream about abundance, and loyalty to an absurd film is a small rejection of common reality.

When films champion wildness, they help us to see our domestic arrangements in ways that fit with our culture's longtime fantasies. Norman Jewison's *Moonstruck* (1987) is a nice place to see this acted out. Written by John Patrick Shanley (Pulitzer Prize winner for a brilliant play with a great title, *Doubt*), the film enacts our conviction that only dangerous love is real love. Lorretta—Cher—had true love once, but her husband was killed by a bus, and after that she was alone for a long time. Then she dated mild, doughy Johnny. When Loretta tells her mother (Olympia Dukakis) that Johnny has proposed and she has accepted, her mother asks if she loves him. Loretta says no, and Mom responds, "Good. When you love them, they drive you crazy because they know they can." That this is the essential quality of true love is acted out in four wolf scenes. First there's the liquor store at the beginning of the film, when the older couple behind the counter argue: Lotte accuses Irv of looking at a some woman "like a wolf." Irv teases his wife that she's never even seen a wolf, and Lotte retorts, "I seen a wolf in everybody I ever met, and I see a wolf in you." Irv gets us out of this by saying slyly, "You know what I see in you, Lotte?" "What?" she asks. "The girl I married." Everybody smiles. We never see Irv and Lotte again. They are there only to announce that every one of us has a wolf inside.

In the next wolf scene, Loretta and Nicholas Cage's Ronny (the sexy brother of her lumpy fiancé) famously tell each other their pasts. This includes a reference to Ronny's ex, who left him when he lost part of his hand in a bread slicer:

> Loretta: You tell me a story and you think you know what it means, but I see what the true story is, and you can't. She didn't leave you! You can't see what you are. I can see everything. You are a wolf!
> Ronny: I'm a wolf?
> Loretta: The big part of you has no words and it's—a wolf.

As he carries her to the bed, she tells him to ravage her so completely that there is nothing left for his brother to marry. And he agrees: "There will be nothing left." The third wolf comes when she scolds Ronny for betraying his brother, Loretta's fiancé. Ronny hotly retorts: "You tell me my life? I'll tell you yours. I'm a wolf? You run to the wolf in me, that don't make you no lamb!" So she's a wolf, too. In the last wolf scene, Ronny and Loretta have been to the opera as a last date, and they are walking. She still plans to marry Johnny. At the opera, they had bumped into Loretta's father with another woman. We are sure by now that the mother preaches the wisdom of marrying a tame friend not because she did so, but because she did not, and suffers for it.

After the opera, Loretta asks to be taken home, but Ronny has them meander to the doorstep of his apartment. Noticing where they are, she says she must leave, and it is his last chance to convince her. Loretta's ambivalence about switching to Ronny is not just about the social embarrassment of canceling an engagement and telling people you are in love with the ex-groom's one-handed brother. After her parents' example, the death of Loretta's first husband only confirmed that true love will clobber you. In the original *Moonstruck* script, the speech ends like this:

> Come upstairs with me, baby! Don't try to live your life out to somebody else's idea of sweet happiness. Don't try to live on milk and cookies when what you want is meat! Red meat just like me! It's wolves run with wolves and nothing else! You're a wolf just like me! Come upstairs with me and get in my bed! Come on! Come on! Come on!

It is uncanny how powerful the wolf theme is, as a modern marker of authenticity. This is the speech Cage delivers in the film:

> Loretta, I love you. Not like they told you love is, and I didn't know this either, but love don't make things nice—it ruins everything. It breaks your heart. It makes things a mess. We aren't here to make things perfect. The snowflakes are perfect. The stars are perfect. Not us. Not us! We are here to ruin ourselves and to break our hearts and love the wrong people and *die*. The storybooks are *bullshit*. Now I want you to come upstairs with me and *get* in my bed!

At first she rolls her eyes a little, but she is persuaded, and so are we. We have almost no opportunities for wildness outside of our private homes, and even there the only place we still use animal metaphors in a positive way is the bedroom. The wolf in the bedroom is key to who we are. What gets Loretta to give a new answer to her mother's question ("Ma, I love him awful") is that Ronny says the storybooks are bullshit; that we should not evade the wolf and settle down with the gentle woodsman or the virginal Red Riding Hood, as the case may be. Instead recognize yourself as the wolf, and when you see another one, run out to meet it and go bay at the moon. Romantic love is imagined as sufficiently intimate and wild as to make up for all the intimacy and wildness that once filled the town. What Ronny offers Loretta appears to be a random, desperate plea; but look again. The speech is not an argument for the thing he is offering, but rather the speech is itself what he is offering: here it is, my passion; I can get us this high, this wild. His audience is not to be convinced so much as impressed.

So here is this ferocious animalism, and willingness to be seen out of control, to be seen barking, as it were; but this wildness is tightly contained in the romantic couple. Yet we make films about this wild intimacy and we all go see the films together, in perfect decorum. Any movie theater anywhere has a culture for you to negotiate—about where to sit in relation to other people, about what to eat and how, about silence or inflated laughter, and on top of that sense of shared space bonding, the whole group is told an emotional story in the dark. That is different from seeing the film at home. You may prefer either, but note the festival qualities of the former. All that said, no matter how good the movies are, you have a better chance of happiness if you do not let actors do all the dressing up in your life while you do all the watching. Even within films, characters dress up often; in between the second and third wolf scene, Loretta got a makeover at the Cinderella Salon.

Do not underestimate the power of costume. When authority is clearly invested in a role, be it king or pediatrician, half the job of ruling or doctoring is done; and for subject or patient, half the work of submitting is done, too. Switching outfits gives you the feeling of the other side. There is pleasure in a false beard or woman's makeup, a peasant's rags or a cleric's robes. Wear a beard or makeup every day and it may lose some of its thrill, but that can be rejuvenated by seeing someone else fetishize it, overdo it, and lavish their psychic attention on its strange human pleasure. Once we were all children, and none of us had beards or breasts. Our adult lives are lived in the costume that biology bloomed out of us, a costuming from within, and we do whatever we may do with these strange and, at first, astonishing and burdensome costumes. Perhaps it is also a cathectic act as people dressed as one another's gender or type knit the group together.

Throughout history people have talked about creating community ritual. Though many holidays seem to have arisen in the fog of time, some were planned out in the hope of satisfying some need in the population. Hanukkah is a good example, as is Memorial Day. In the nineteenth century, many people proposed secular cults to replace the community ritual that religion once provided. The Freemasons were all about this: consider your dollar bill. In the French Revolution, the Committee of Public Safety cooked up all sorts of festivals, borrowing imagery from democratic ancient Athens and the Roman republic. The best new holiday is Gay Pride. Its parade is political, but also has carnival and costume, and its festivities are celebrated as a weekend, while most celebrations try for only one day. What becomes clear is that it is not enough to come out of the closet; you have to also leave the house. Today it would feel silly to call for a new holiday, like *Seinfeld's* Frank Costanza, who invented Festivus: "a holiday for the rest of us." It is a nice rhyme, but it is comedy—precisely because new holidays are not necessary. More seriously, we need to think about how community celebration works in our lives and how we might make use of the gatherings we already have. It seems worth considering how we spend our time, with some new criteria.

Look at a calendar. See what celebration is coming up. Ask yourself these nine questions:

1. Is it going to give you a chance to act in an unusually free way in public, perhaps because of costume, darkness, or ritual?
2. Might you dance with abandon?
3. Does it recite and enact a dramatic story?
4. Is there an element of topsy-turvy, of power and gender inversion?
5. Is there absurdity?
6. Is there special food abundance?
7. Is there a crowd?
8. Is there nakedness?
9. Does the celebration have anything to do with you?

With even three yes answers, you should go. Maybe it sounds like these questions are describing a bacchanal, but even Thanksgiving invites you to five, with its special food abundance; its weird morality tale about sharing with people you eventually annihilate; its Macy's parade—gargantuan, freakish, absurd balloons; the parade's great crowd; and finally, if you live in the United States, with the celebration's relevance to you. The Jones Beach July 4th fireworks gives you 1, 3, 7, 8, and 9. I provided a lot of details about what people did at ancient festivals and medieval carnivals because I wanted you to be able to note whether you were envious of any of it, and to be as specific as possible. Say you were envious of bathing with a piglet in the Mediterranean Sea, surrounded by all the nonvirgin women of your world, and of later, in a controlled, ritualized frenzy, killing the piglet and offering it to the gods in hopes of getting a good birth in exchange. Or say you were envious of the husband-and-wife partying that went on at the Eleusinian mysteries. The reader tempted by the one may recoil at the other, but we can all understand the temptation given how tough it is to be a human being in any age, how hard it can be to get pregnant, how hard to have kids, how hard to raise them. The constant companion of anything having to do with your kids, the ones you have or the ones you wish to have, is a torrent of hope. We may also speak of a torrent of hope in our desire to be fully alive, erotic, and in love. Other moments in history have offered so many rituals to make sure these torrents had some theater of action. At the movies we watch lots of people in costume, and it must do us some good; perhaps it reminds us how our roles and personalities could theoretically be taken away or put on.

In the past, people believed that participation in the ritual or the parade actually kept the community safe, magically. We don't have magic nowadays, but the belief is no less true. If we go to the Halloween parade, the Halloween parade continues to exist, with people like us in it, whatever that might mean. I believe such activity repairs the walls of your mental world. If you only think about it, watch it on television, or read about it in the paper, it is not enough. If you leave out the part where you actually show up with hundreds or thousands of other people, you've lost an immeasurable, indescribable part of the experience. There are only a few pragmatic routes to happiness, and celebration is one of them. Get out there.

Conclusion

The Triumph of Experience

In all circles of modern life, no one seems shy about using the phrase "It turns out that ..." But we do not live at the end of history, and we are not boring perfection, too normal to be expected to change. "It turns out that ..." is a generally inappropriate phrase. We are just another era, in another set of deeply integrated nonsense. We feel like we live in a disenchanted world. The list of creatures in whom nobody believes anymore is impressive: vampires, witches, ghosts, and demons; in some circles, aliens; and, in some circles, God. Though we feel proud of how much we know nowadays, we can find ourselves sad that our world is more reasonable than magical. Cheer up: we don't know as much as we think we do, and we are, in fact, magical. That is the correction I have been trying to make. I hope this book brought into vision the enchanted world: the enchanted details of our own particular moment's beliefs and habits—our diets as internal corsets, our conviction that the well-fed are hungry; but also the enchantment that we share with so much of Western history—our wolves, our missing girls, our baths, our passages through hell and back. This is who we are. We are part of something rich and strange. To be aware of it is to consciously take part in it and to be happy, or happy enough. We do not need to hide in fear, whether in the form of religion or nationalism, scientism or politics. Life—this wolfless, literate, pharmaceutical, big-screen fantasy that we share—ought to be embraced for what it is: a victory of desires, the dream of our ancestors, and another tumble in the kaleidoscope of historical culture. Eat it raw or eat it cooked, but feast, and go out there and meet your wolves, let yourself be feasted on, see what happens.

Our culture of what to do about happiness is full of double meanings, cheating, and secrets, because we are using the whole thing as a distraction. If we want to think clearly, with room for originality, we must notice that being obsessed with the stuff you do about happiness is a wrong turn, a terrible tangle of double talk and contrary information about how to make a happy life. We need to relax these polemics and try on different variations of behavior. Maybe it is okay to be fat. Maybe drugs are all both good and bad and you should try out a new relationship with them to see what makes you happy. Maybe you want to rethink whether what you get out of shopping has anything to do with buying a lot. Maybe common concerns about people who have a lot of sex or no sex are just noise and things are okay the way they are for now. Maybe we can stop feeling so conflicted about shallow American culture and recognize that we are lucky to have something shallow to share. Why not put your arms in the air and party like you just don't care? You are a mammal with extraordinary potential, but we have to *take care of you* if you are going to fulfill that potential. You have to do some work—wisdom work, celebration work—and you also have to learn to be a truth detector, to know that the rules for happiness propagated by the culture at large are not to be allowed to take up too much time and energy. Wrote Montaigne, "The reason why we doubt hardly anything is that we never test our common impressions. We do not probe the base, where the fault and weakness lies; we dispute only about the branches."[1] He asked, "What am I to make of a virtue that I saw in credit yesterday, that will be discredited tomorrow, and that becomes a crime on the other side of the river? What of a truth that is bounded by these mountains and is falsehood to the world that lives beyond?"[2] When it comes to happiness drugs and health, and happiness regimes, even the most scientific, I'd say go with skeptical fidelity at the very most, and maybe relax your vigilance or your guilt. Probably our belief in these laws and standards is unwarranted. By contrast, there are some things that *will* make you happy.

As I discussed in the introduction to this book, happiness has not increased since 1950. Plotted on a graph, all sorts of measures of living conditions climb in broad, steady strokes, but happiness just lies there. What if we built a situation that could be the best civilization ever but we could not quite function in it? What if we are finally rich enough (in some circles) and wise enough (in some circles) to look after all of society properly but we just cannot get it together and make it happen? The great meaning of the world is each individual person getting through life with some happiness. It is time to work out the ambivalence and complexities of how we feel about our abundance and then learn to make use of what we find, for the sake of our own happiness and the happiness of other people. Being terrified of all sorts of dangers makes it easy for us to hide from all our ambivalence, but for most of us, most often, being terrified is a waste of time.

We are tempted to assume that the difference in money-and-happiness ratios between the 1950s and the 2000s can be explained by concluding that money doesn't buy happiness. But that's not true: money can buy happiness. Yet we feel guilty and worried over our abundance. We ruin it all with anxieties. Anxieties are biological memories of the wolf. In the absence of the wolf, we attach the anxiety to new things. We have to cut this out. Having failed to improve from 1950 to 2000, are we going to make any happiness progress from the 2000s to 2050? What would progress look like? What would make more of us report ourselves happy with more aspects of our lives? It seems to me that to make progress we will have to be open to suggestions from science, but also much more mindful of what people actually do. Tomorrow's happiness advice will be different from today's. It won't just tell you to do something weird in order to live longer; it will no longer be so focused on living longer. That is just how it is. Culture changes. For a while, one way we dealt with mortality was to fixate on getting a statue of yourself put up somewhere after you died. We don't even think about statues anymore, not like that. Right now, it's all longevity. Later, it will be something else. The changing cultural concerns of future generations will lead them to different kinds of scientific questions and answers, and those will lead to different advice. The only way to make progress is to get your head out of your century and look around. If we remember the history lessons I have pointed to in this book, it is possible that we can make some modest sour-charm gains. The idea is not to abandon the search for happiness but to be suspicious of the same old ways of thinking about it. We should scientifically attack the question with glee and abandon, take our results lightly, and not get seduced into feeling shame or guilt over what we are not managing to do in the service of our own happiness.

The reason we cannot do everything we want to do in order to be happy is that the three kinds of happiness conflict with one another. If you are working hard, you are not relaxing; both are good goals, and you need a chart to know when to do each. Most of us just use the standard method of the forty-hour work week and the calendar of holidays and vacations. But think of all the other happiness acts that are in conflict and for which we have no standard schedule. You may want cake and a lean body; but at any given moment, which one of these should you be seeking? As I see it, this realization is almost enough to start making decisions that feel more informed, more processed through the intellect. But it is a good idea to think about how you want to apportion your three kinds of happiness.

It seems worth the effort to make a few lists of things you can do to make yourself happy and try to sort them. Consider a version of the three main categories of happiness. The following lists are of course meant for adults, and we each have to avoid some activities because they don't work well for our safety or our relationships:

Good-Day Happiness
What Makes a Happy Day for You?
Seeing friends
Chatting with neighbors
Eating chocolate cake
Having a few drinks
Playing with your kids
Reading a good book
Waking early for a relaxing morning
Taking a walk
Playing a game
Solving a puzzle
Playing a sport

Taking care of someone sweet
Shopping
Bathing
Getting a massage
Having sex
Working with your hands
Watching entertainment
Cooking

Euphoria
How Do You Get Euphoria?
Great sex
Music
Meditation
Drugs
Crowd celebration
Dance
Dangerous sports
Art

A Happy Life
*What Do You Need to Have, or Be Working Toward, in Order to Like
Your Life?*
Family
Friendships
Celebrations and rituals
Travel
Study
Skills mastered
Money in the bank
Community service
An attractive appearance
Adventure
Serving as an inspiration
A history of a lot of good days
A history of some euphoria

These lists will vary a great deal from person to person, but if you have an idea of what your lists look like, you can have some idea of how you might want to make changes—either in what you do or in how you think about it.

Don't overschedule. Most people can manage to attend maybe one parade a year. If you try for both a Thanksgiving Day parade and a Halloween parade, it is likely that for one of them, you or your spouse or child will not be feeling good, or it will be cold and rainy, or something else will get in your way. But if you *try* for both every year, most years you will get to at least one. It is tedious to arrive early for Fourth of July fireworks, and get a good spot, and wait; and if you have to drive there, it will be tedious to get out of the parking lot afterward. But while you are there you get to feel the vibrations of the noise, maybe listen to the broadcast of American music over the radio, and see something exceedingly beautiful and strange with your fellows. Everyone will say "Ahh" and "Oooh," and then they will laugh at their part in the great chorus of ahhs and oohs. Try, once in a while, to get right under the action.

If you are hoping to make room in the day for some hard work, you probably will not be able to do very many good-day things: either talk to a friend or read a chapter in a book; either take a walk or see a movie. You decide for yourself how often you eat cake. If that is every day, you may have to live with the consequence of not being the height of modern attractiveness. If you want to stay up late, you are going to have to feel groggy much of the next day. It might as well be a conscious choice.

It is worth hashing out what kind of choices we can make. Think about euphoria: When have you had it? How much of your euphoria is public and how much private? How long do you usually go between events where you get a little large-group elation? What would happen if you shortened the cycle? Maybe it wouldn't be worth the effort, but maybe it would. If you go shopping, remember what a magical and odd place you are visiting. Think about what different visions of abundance do to you. Do you want to bring everything home? Why would all that stuff be better stored at your house? This is the marketplace gone to paradise. If you do not like to be in stores, go to a big store or a cluster of them with hours to kill and nothing you need to purchase, and walk around and see what it feels like to witness the abundance. If you want, buy something. Consider the pleasure of anticipation and of purchase.

Could you throw a big party once every six months? Once every two years? Give blood twice a year? If it has been a while, make a point of going into a flower shop or a beautiful library. How many concerts or other live music events did you get to last year? Last decade? Pick a big show coming up and buy the tickets. Go to a local place and hear a good band. Compare how the two make you feel. Some of us do not see much art and some of us do not see much nature. Go see a game, or go see two, one in a giant arena, one at the local park. Try karaoke. Go to a fashionable crowd and watch the people. Go to the theater. Find different ways to have a good soak. If you rarely go somewhere to get groomed, go get groomed. If you rarely go hear a lecture, go hear one. It is likely you have tried almost all of these, of course, and you have good reasons that you do not do them, but consider trying them again and thinking about the way they connect you to other people, and see how that makes you feel. Remember just this: when it comes to enjoying a parade, you have to show up. Like costume, you cannot just think about any of this. You have to try it on.

Ours is a culture that makes a distinction between behaviors that bring true happiness and behaviors that only make you *feel* happy! Yes, Henry VIII must have had a painful inner hunger and a deep loneliness, but he also had a turkey leg in one hand and six nubile queens, successively, in the other. It is almost magical, maniacal, that we concentrate all of our intellectual and mature analysis on one response to Henry and other "oversexed" gourmands: that they are actually less happy, and less fulfilled, than the norm. It is even more remarkable when you notice how much our culture fans the flames of desire for both a nice, big, greasy hunk of bird meat and also a procession of hot Anne Boleyns.

Today, we see fewer summer or year-round communities based on an ideal regimen for health and happiness. Yet Americans have worlds of communities that are primarily organized around quitting something as a route to happiness. Why does happiness seem to us to be dependent on active, aggressive self-control and self-denial? Throughout history people have made a performance of self-control and self-denial, for beauty or for God; but denial is not usually this exquisitely wrought. We even legislate these denials on an unprecedented scale. There used to be laws against saying all sorts of things, but you could eat, drink, and smoke what you wanted, at any age, almost anywhere. Americans are notably legalistic in such matters; there are other cultures that avoid tragic indulgence without making a cultural movement—let alone a federal case—out of it. Still, across cultures, a key route to happiness is periodic devotion to controlling one's physical desires.

If you want to be happy, do some experiments. Try a variety of ways of being with people. Go to the Irish pub where they sing drinking songs, and sing along. At weddings, get up and dance. Go to the Coney Island Mermaid Parade, preferably dressed up like a mermaid. (You might not want to bring your parents, as there are always some mermaids who are pretty much naked.) If you see a bunch of people standing around a guitar player in Washington Square Park, and everyone is singing "Hey Jude," you go over and sing, too. Go to big festivals, too: sports arenas, parades, music in the park. Join in when everyone is singing along or doing a chant. Think about how you feel in such settings. Speak clearly to yourself about what you get from the experience, and what you don't get; which gatherings feel cheap or empty and which feel full of value. Rethink your drugs. Find out if coffee makes you happy. If you are a person who drinks a lot of alcohol sometimes, think of the philosophers who claimed that drink is the workingman's access to the spiritual. Drinking too much will kill you. You shouldn't do it. But as a culture, must we pretend people drink only to anesthetize themselves or to have a stupid kind of fun? Drinking is also *interesting*.

Sort the hard news from the soft news, and think about the kinds of narratives that get told in the soft news. Think about the stories that interest you. Ask yourself which of the upsetting stories make you happy. Are you drawn to the vicarious experience of any particular kind of loss? Are the news stories that draw your interest depressing you and not acting like myth? If so, stop reading them. But you may discover that these bloody family dramas do in fact act like myth. Recognizing this doesn't do any harm, just as showing up to a large-scale parade in part as an emotional detective does not do that experience any harm. Both are like drugs: if you want to examine your response to scotch, you are still going to get drunk if you drink enough. With the parade, you have to show up and relax and let yourself be part of it. You will not have to try hard. Think about how you share the news with friends and strangers.

The poets, not a giddy bunch, have left us evidence that life will bloom into happiness now and again, for no reason. Happiness, like sadness, sometimes arrives of its own accord. To be treated to these mysterious waves of happiness, all we have to do is be there when they happen, live to see them. True, coffee, tea, or wine may have been on hand, so it might be a good idea to get a stimulating beverage. But mostly all you have to do is hang on and wait. Just as you don't have to do anything to help make spring come, the winged happiness that the poets speak of will come on its own. It is coming. Earth will swing on its ellipse whatever you do, and crocuses will come up. The poets say happiness comes like this, though we cannot see the works. It is coming. This is an extremely encouraging insight, and I hope I will be able to remember it.

This book has also addressed the matter of truth for its own sake—not to do with happiness, but with reality. Consider a whole century of men and women straining to conserve the body's energy, minimizing sport and exertion in order not to overspend their reserves, and then the entire next century straining to exercise the body so that it will become more efficient. You have been told by physicists and yogis that reality is not what it seems, that your mind makes the world you live in, and you believe it; but you also don't believe it. Half the point of this book was for me to try to cheer everyone up at once. The other half was to demonstrate ways that we look up at the blue sky and say "green." What I have offered is, in its own way, a philosophy. I have tried to show the disjuncture between what we do and what we say we are doing. I hope I have marshaled the evidence necessary to show what a vague hold on reality we have.

The happiness myth makes us spend our time and energy trying to be happy through our historical moment's big ideas, and failing just enough to keep us fixated on the problems as stated. What should we be doing instead? The philosophers and wisdom writers throughout history have told us what we should do in life if we can get past our usual problems (pain, fear, narcissism). I offered the list of their suggestions, with attribution, at the start of the first section of this book, titled "Wisdom." Here I will just give a swift list of activities so good that they need no arguments or fanfare. They are: being loving to your spouse, nurturing your children, tending to your extended family, nurturing friendships, helping local strangers, helping strangers far away, caring for animals, engaging in fine art and the arts of living (poetry, prose, painting, sculpture, music, dance, architecture, cooking, entertaining, gardening, decor), risking both being in the world and keeping apart, doing philosophy, learning the art of traveling and the art of staying home, planning for the future of humanity, and increasing the world's knowledge. Our cultural nonsense about the exact proportions of the body and the superiority of productivity over euphoria—this stuff cuts into our ability to enact the sublime through practicing the good things. The changing cultural rules give us a way to be worried about something that is within our control, and we forget that it is relatively meaningless. Also, this cultural advice limits your choices within the perfectly reasonable realm of possibilities, so that, for instance, nowadays women spend time grooming together, and men don't. Maybe it would be better for the guys if they did. Consider the community function of the old barber shop. If you live at the wrong time, you have to go out of your way to get certain things. It is worth paying attention.

I offer this book as a crowbar to help separate us from our historical moment so we can get a little height and view the subject with less distortion. It is not just a matter of distortion due to vantage point; there is also distortion due to remarkable temporal prejudice. You may love the Renaissance, but you wouldn't want to use their toilet paper. Or take their doctors' advice. Or believe what they believed about time and the universe. Or raise your children the way they raised theirs. Or really believe them about anything. But how can we think we are so great when we have almost no vision of hope for the future other than that we will have lost ten pounds and that a girl, one person out of 300 million Americans, is missing and we hope she will turn up alive. We want to understand the context of these hopes. And knowing the context should allow us to be more efficient in our experience of these hopes, so that we are free to hope for more. We start by working to shake off the myth of knowing. It is especially good to shake off the myth of knowing what we are supposed to be doing. Then, in this less certain state, start sketching out your happiness lists. Start with writing things you actually do; then make additions to each list, noting what you might like to add to your gallery of daily-happiness-type pleasures, ecstasies, and contributions toward lifelong happiness.

Be alert to the convictions of the experts—who, after all, have to say something. Think about which types of experts have changed their advice in your lifetime, and how frequently. When someone makes an announcement about what you need to do, if the advice category seems changeable, be more suspicious; demand more evidence before you change your behavior. Or try the new advice, but see yourself in the context of history, "trying something," as so many people have tried things before. Do it when (or where) other people are doing it, and you get an extra lift. But do not let it be your new anxiety. It is not serious enough for that; it is not real enough for that. Present-day convictions about what normal people do are much too certain. What people eat, what they wear, how they relax, how they party, whether they have sex and how they feel about it, the drugs they take or refuse—it is all tremendously dependent on time and place, and it is not all the same. It is a lottery. If you live in an era that has encouraged you to consider only half of the items on each of your happiness lists, it is up to you to make an effort and do some experiments on yourself. I emphatically do not mean that you can go off on your own morally. I'm talking about things many of us just don't do enough of. Try a parade, a spa, a roller coaster. Talk to neighbors. Do something in the community and for the community. Go to a show. Go to a crowded beach and go to an empty one. Inspire a young person. Go swim in water that assaults you and go practice your back float in a still lake. Go to a local celebration like a saint's feast or a parking-lot carnival. Ask yourself how these feel, and take some notes on the answer. While you are there, and a week later, ask yourself if you are glad you chose to try this, what was the most negative thing about doing it, and how often you would like to do it in the course of a week or a year or a decade.

Cosmopolitan eras like ours are lucky. We have the opportunity to be one of the more savvy periods in history, to pull back from our convictions, hold still, and take in the great variety of outlooks. When someone says that "they" have now got something figured out, you may say aloud or in your head, "No, *they* probably don't." People are shouting too many philosophies of health and happiness at us, and we do not need to sort through the cacophony, but only to notice the character of concerns. We seem obsessed with motivation, rallying ourselves to something beyond the life available to us right now, and we treat this motivation as if it were a major part of the history of wisdom, which it is not. We seem fanatical for longevity—which I do not dismiss entirely, of course, but which seems painfully at odds with people's actual behavior. We try to promote sobriety, but again our actions do not accord with our words, and that will not square until we start respecting the positive aspects of drugs. We are obsessed with productive decorum, and require it everywhere but in art, today's haven of topsy-turvy and costume. We are still acting out our victory over the primary battles of life: after ten thousand years of farming grain, fabricating clothes, and outsmarting wolves, we don't have to do any of that anymore. These serve now as our great symbols of authenticity. We need to awaken from not noticing them, from taking them as given. As we have seen, there are other ways to see things.

Acknowledgments

I am grateful for conversations I have had with Jeffrey Ringle and Amy Allison Hecht. I am also happy to thank Carolyn Eisen Hecht for reading the manuscript and offering immense encouragement. Tanya Elder and Chris Krol helped too. John Chaneski often disagrees with me and frequently changes my mind. Along different lines, I have lately realized that I never mention Michel Foucault in this book, but I cannot imagine writing it had I not read, over a decade ago, his *History of Sexuality*, *Madness and Civilization*, and *The Order of Things*. Thanks to Jim Rutman at Sterling Lord Literistic. I can't think of anything the Keller-Swansons did to help; indeed, they frequently interrupted me, for which I am grateful. Not least, thanks to the great people at Harper-SanFrancisco, especially editorial director Michael Maudlin and my editor, Roger Freet, for much helpful direction, executive managing editor, Terri Leonard, for getting the manuscript to press, and publicist, Julie Mitchell, and assistant editor, Kris Ashley for their help too. Much thanks to designers Joseph Rutt, Joan Olson, and Jim Warner for an especially delicious-looking book. For all that is strange, ridiculous, or wrong in the manuscript, the fault is all my own.

Notes

WISDOM

1. Montaigne's subject here was why he was sticking with Catholicism while much of Europe at large was seduced by the "new knowledge" of Protestantism. Protestants argued that much of Catholicism was just an accumulation of mumbo-jumbo developed since Jesus. Montaigne's position was that religion is not one of those things you can get right, it is about a completely mysterious something—if it is about anything more than custom. So why create upheaval and bloodshed over its details? Better to uphold the religion in which you were raised. The Catholic Church needed the support and promoted Montaigne's skeptical fidelity for a century.

2. Michel de Montaigne, *The Complete Essays of Montaigne*, trans. Donald M. Frame (Stanford, CA: Stanford Univ. Press, 1958), 358–59.

3. Montaigne, *Complete Essays*, 359.

4. Bertrand Russell, *The Conquest of Happiness* (New York: Liveright, 1930), 153.

CHAPTER 1: KNOW YOURSELF

1. Marcus Aurelius, *Meditations* (New York: Dover, 1997), 79.

2. Benedict Spinoza, *Ethics* (New York: Hafner, 1949), 227.

3. Carl Jung, "Paracelsus as a Spiritual Phenomenon," in *Alchemical Studies, Collected Works* (Princeton: Princeton Univ. Press, 1983), 13:109–89.

4. Aurelius, *Meditations*, 20; Shakespeare, *Hamlet*, act 3, scene 2.

5. Aurelius, *Meditations*, 19.

6. Elizabeth Cady Stanton, "The Solitude of Self," in *History of Woman Suffrage*, ed. Stanton et al. (New York, 1902), 4:189–91.

7. Montaigne, *Complete Essays*, 365.

8. Russell, 113–14, 123. UNESCO Institute for Statistics, *Regional Adult Illiteracy Rate and Population by Gender*, September 2006 assessment, unesco.org. UNESCO began keeping track in 1950, calling adult literacy the ability of those 15 years old or older to read and write a simple note in any language. The U.S. Department of Education's National Center for Education Statistics estimates U.S. illiteracy for 1870 at 20 percent; for 1930 at 4 percent; and 1979 at 0.6 percent. Of course race and gender matter: in 1930 the illiteracy rate for "Black and Others" was 16 percent.

CHAPTER 2: CONTROL YOUR DESIRES

1. "The Discourse of the Teaching Bequeathed by the Buddha (Just Before His Parinibbana)" was translated into the Chinese, from which it survives, by the Indian scholar Acarya Kumarajiva sometime prior to the year 956 of the Buddhist Era, around 344–413 C.E.

2. *The Vatican Sayings*, http://www.epicurus.info/etexts/VS.html. This is a collection of maxims, titled "The Sayings of Epicurus," that was rediscovered in 1888 within a fourteenth-century Vatican manuscript that also contained Marcus Aurelius's *Meditations* and Epictetus's *Manual*.

3. Aurelius, *Meditations*, 20.

4. Aurelius, *Meditations*, 63.

5. Aurelius, *Meditations*, 89–90.

6. Russell, *Conquest of Happiness*, 91.

7. Russell, *Conquest of Happiness*, 91.

8. Spinoza, *Ethics*, 223.

9. Montaigne, "Of Experience," in *The Complete Essays of Montaigne*, trans. Donald M. Frame (Stanford, CA: Stanford Univ. Press, 1958), 854.

10. Montaigne, *Complete Essays*, 855.

11. See James Atlas, "The Fall of Fun," *New Yorker*, November 18, 1996; and Mark Kingwell, *In Pursuit of Happiness: Better Living from Plato to Prozac* (New York: Crown, 2000), 115.

CHAPTER 3: TAKE WHAT'S YOURS

1. Aurelius, *Meditations*, 98.

2. Aurelius, *Meditations*, 85.

3. For a juicy recent portrait of Alcibiades, see Lucy Hughes-Hallett, *Heroes* (New York: Knopf, 2005).

4. George Bernard Shaw, "Epistle Dedicatory," in *Man and Superman*, http://www.4literature.net/George_Bernard_Shaw/Man_and_Superman/8.html.

5. Ralph Waldo Emerson, *Essays* (Cambridge, MA: Riverside Press, 1883), 1:48.

6. Emerson, *Essays*, 1:69.

7. Emerson, *Essays*, 1:69.

8. William James, *The Varieties of Religious Experience* (New York: New American Library), 88.

CHAPTER 4: REMEMBER DEATH

1. Geoffrey Gorer, "The Pornography of Death," *Encounter* 5 (1955): 50–51.

2. Aurelius, *Meditations*, 93.

3. *L'Intransigeant*, 1922. I take the anecdote from Alain de Botton's sweet and useful *How Proust Can Save Your Life* (New York: Vintage, 1998), 6–8.

4. "Discourse of the Teaching."

5. Shaw, "Epistle Dedicatory."

6. Montaigne, *Complete Essays*, 361–62.

7. Heidegger broke bread with Fascists; Schopenhauer pushed a woman down the stairs (for talking in the hall) with such violence that he was ordered by a court to pay her an annuity until she died, many years later. Wittgenstein was born wealthy but banished himself to grim poverty. Simone Weil went on a hunger strike in solidarity with those in the camps in World War II, such that her tuberculosis killed her. Karl Marx failed to support his family to the point of perhaps fatal lack of food and medicine, Plato said awful things about the common laborer, Jefferson had slaves, and Nietzsche wrote vile things about the various "races." As for the sexists, take your pick.

8. John Stuart Mill, *Utilitarianism* (New York: Barnes and Noble, 2005), 11.

9. Epicurus, "Letter to Menoecus," in *The Essential Epicurus*, trans. Eugene O'Connor (Amherst, NY: Prometheus Books, 1993), 61.

DRUGS

1. See the Center for Disease Control Web site, under the National Center for Health Statistics: http://www.cdc.gov/nchs/hus.htm

CHAPTER 5: WHAT MAKES A GOOD DRUG BAD

1. Lu Yu's book *Ch'a ching*, or *Tea Classic*, of 780 C.E. treated every subject related to tea, from growing plants to brewing and drinking, including a detailed description of a formal tea ceremony using twenty-seven pieces of equipment.

2. Samuel Johnson, review of *A Journal of Eight Days' Journey*, *Literary Magazine* 2, no. 13 (London, 1757).

3. The British Tea Council is an independent body dedicated to promoting tea drinking.

4. Richard Davenport-Hines, *The Pursuit of Oblivion: A Global History of Narcotics* (New York: Norton, 2002), 190–91.

5. Consider one historian's colorful description: "This process was speeded up by eating—out of sheer necessity—mushrooms, toadstools, and grasses of all kinds. It must have been incredible, the sight of an entire community betaking itself to the poppy fields for want of anything better to eat. Such behavior was reported in early-modern Italy, where people also deliberately sniffed salves and lotions in the hope of sailing away on a cloud of bliss. In the rest of Europe as well, there had been reports in preceding centuries of bread made of poppy extract and even of hemp-seed flour being used to make bread dough." Herman Pleij, *Dreaming of Cockaigne: Medieval*

Fantasies of the Perfect Life, trans. Diane Webb (New York: Columbia Univ. Press, 2001), 125.

6. Davenport-Hines, *Pursuit of Oblivion*, 38.

7. As a lieutenant governor of Bengal, Sir Charles Elliott explained in 1892 that it was as "reasonable to suppose that excessive ganja smoking may be due to insanity, as that insanity may be due to excessive ganja smoking." Davenport-Hines, *Pursuit of Oblivion*, 191.

8. Théodule Ribot, *Diseases of the Will*, trans. Merwin-Marie Snell (Chicago, 1896), 38.

CHAPTER 6: COCAINE AND OPIUM

1. Freud to Martha, June 2, 1884, in Ernest Jones, *The Life and Work of Sigmund Freud* (New York: Basic Books, 1953), 84.

2. See Dominic Streatfield, *Cocaine* (New York: Picador, 2001), 76–77.

3. Davenport-Hines, *Pursuit of Oblivion*, 205.

4. W. Oscar Jennings, *The Morphia Habit and Its Voluntary Renunciation: A Personal Relation of a Suppression After 25 Years of Addiction* (London, 1909), 63.

5. Davenport-Hines, *Pursuit of Oblivion*, 213.

6. Davenport-Hines, *Pursuit of Oblivion*, 214.

7. Thomas S. Blair, "The Relation of Drug Addiction to Industry," *Journal of Industrial Hygiene* 1 (1919): 295.

8. J. M. Scott, *The White Poppy* (New York: Funk & Wagnells, 1969), 5, 46–82, 109–25.

9. Galen, *Souvenirs d'un médecin*, trans. Paul Moraux (Paris: Société d'édition "Les Belles Lettres," 1985), 134–39.

10. T. W. Africa, "The Opium Addiction of Marcus Aurelius," *Journal of the History of Ideas* 22 (1961): 97–102.

11. Edward Gibbon, *The History of the Decline and Fall of the Roman Empire*, vol. 1 (1776), chap. 3.

12. For more on this, see Barbara Hodgson, *Into the Arms of Morpheus: The Tragic History of Laudanum, Morphine, and Patent Medicines* (Buffalo, NY: Firefly, 2001), 22.

13. Davenport-Hines, *Pursuit of Oblivion*, 51.

14. Charles-Louis de Secondat, baron de La Brède et de Montesquieu, *The Persian Letters* (Indianapolis, IN: Hackett Publishing, 1999), 57–58.

15. Thomas De Quincey, *Confessions of an English Opium-Eater; and Other Writings* (New York: Oxford Univ. Press, 1998), 74.

16. As cited in Hodgson, *Into the Arms of Morpheus*, 45.

17. I've taken the liberty of smoothing out some of the arcane punctuation and spelling in this poem. Cited in full in Hodgson, *Into the Arms of Morpheus*, 65.

18. Wilkie Collins, *Armadale* (Oxford: Oxford Univ. Press, 1999), 513–14.

19. Hodgson, *Into the Arms of Morpheus*, 93.

20. Alphonse Daudet, *L'évangeliste*, as cited in Hodgson, *Into the Arms of Morpheus*, 93.

21. Russell, *Conquest of Happiness*, 51.

CHAPTER 7: RELIGION AND REVELATION

1. If you look at Psalm 104, you find that much is foreign: we do not value oil to make our faces shine; religious leaders mostly do not dismiss the idea of an afterlife for animals; we believe the Earth does move and the sun does not move around us; we believe that the waters *will* come back over the Earth. It has all been made into nonsense, except the part about wine; it still gladdens the heart.

2. James, *Varieties of Religious Experience*, 297.

3. James, *Varieties of Religious Experience*, 298. This discussion continues: "How to regard them is the question,—for they are so discontinuous with ordinary consciousness. Yet they may determine attitudes though they cannot furnish formulas, and open a region though they fail to give a map. At any rate, they forbid a premature closing of our accounts with reality."

4. Aldous Huxley, *"The Doors of Perception" and "Heaven and Hell"* (New York: Harper Perennial, 2004), 22–23.

5. Huxley, *Doors of Perception*, 26.

6. Huxley, *Doors of Perception*, 41.

7. Huxley, *Doors of Perception*, 73.

8. Huxley, *Doors of Perception*, 69.

9. Huxley, *Doors of Perception*, 79.

10. Rick Doblin, "Pahnke's 'Good Friday Experiment': A Long-Term Follow-Up and Methodological Critique," *Journal of Transpersonal Psychology* 23, no. 1 (1991). The full text is available at http://www.druglibrary.org/schaffer/lsd/doblin.htm.

11. Jeanne Malmgren, "The Good Friday Marsh Chapel Experiment; THEN — Rev. Mike Young — NOW 'TUNE IN, TURN ON, GET WELL?,'" *St. Petersburg Times*, November 27, 1994.

12. Mary Barnard, "The God in the Flowerpot," *American Scholar*, Autumn 1963, 584, 586.

13. Henri Bergson had suggested that the Hindus' and Greeks' invention of religion was a "divine rapture" found in intoxicating beverages; Robert Graves and Alan Watts also talked about the origins of all religion being traceable to chemical highs. Recently, Dutch historian Herman Pleij has agreed, saying that the medieval period was experienced in the vivid reverie and generative stupor of all sorts of natural intoxicants: "Where did the penchant for mysticism come from otherwise?" Pleij, *Dreaming of Cockaigne*, 127. I cannot fully agree, as drugless religious bliss exists; but there is clearly something to the idea.

14. Huston Smith, "Do Drugs Have Religious Import?" *Journal of Philosophy* 61, no. 18 (September 17, 1964): 517–30.

15. Smith found the first quote in Willis W. Harman, "The Issue of the Consciousness-Expanding Drugs," *Main Currents in Modern Thought* 20, no. 1 (September–October 1963): 10–11. The second experience was quoted by William James in *The Varieties of Religious Experience* and was attributed by James to a Dr. R. M. Bucke, the author of *Cosmic Consciousness*.

16. Smith, "Do Drugs Have Religious Import?" See: http://www.druglibrary.org/schaffer/lsd/hsmith.htm for a searchable full text of the article.

17. Smith, "Do Drugs Have Religious Import?" 530.

CHAPTER 8: DRUGS TODAY: MUSIC AND SOLACE

1. "Better Than Well," *Economist*, April 6, 1996, 99–100.

2. Julie Holland, *Ecstasy: The Complete Guide: A Comprehensive Look at the Risks and Benefits of MDMA* (Rochester, VT: Park Street Press, 2001).

3. "Using MDMA in Alternative Medicine: An Interview with Andrew Weil," in Holland, *Ecstasy*, 287.

4. Nicholas Saunders, with Rick Doblin, *Ecstasy: Dance, Trance, and Transformation* (Oakland, CA: Quick American Archives, 1996), 115.

5. Saunders, *Ecstasy*, 117.

6. Saunders, *Ecstasy*, 117–18.

7. The first quotation is from the Woody Allen film *Love and Death*. The second is from Arthur Schopenhauer, *The World as Will and Representation*, trans. E. E. J. Payne, 2 vols. (New York: Dover, 1969), 2:586.

MONEY

1. Adam Smith, *The Theory of Moral Sentiments*, ed. Knud Konssen (Cambridge: Cambridge University Press, 2002), 211–14, 216.

2. Frank M. Andrews and Stephen Withey, *Social Indicators of Well-Being: Americans' Perceptions of Life Quality* (New York: Plenum Press, 1976), 332.

3. Joseph Veroff, Elizabeth Douvan, and Richard A. Kulka, *The Inner American: A Self-Portrait from 1957 to 1976* (New York: Basic Books, 1981), 98.

CHAPTER 9: HAPPILY EVER AFTER

1. Robert E. Lane, *The Loss of Happiness in Market Democracies* (New Haven, CT: Yale Univ. Press, 2000).

2. Princeton Research Associates, July 12–15, 1994, cited in *American Enterprise*, November–December 1994, 99.

3. Philip Brickman and Donald T. Campbell, "Hedonic Relativism and Planning the Good Society," in *Adaptation-Level Theory*, ed. M. H. Appley (New York: Academic Press, 1971).

4. Isaac M. Lipkus, Claudia Dalbert, and Ilene C. Siegler, "The Importance of Distinguishing the Belief in a Just World for Self Versus for Others: Implications for Psychology of Well-Being," *Personality and Social Psychology Bulletin* 22 (1996): 666–77.

5. Lane, *Loss of Happiness*, 73.

6. Sara J. Solnick and David Hemenway, "Is More Always Better?: A Survey on Positional Concerns," *Journal of Economic Behavior and Organization* 37, no. 3 (1998): 373–83.

7. D. A Schade and D. Kahneman, "Does Living in California Make People Happy? A Focusing Illusion in Judgments of Life Satisfaction," *Psychological Science* 9, no. 5 (1998): 340–529.

8. Elizabeth W. Dunn, T. D. Wilson, and D. T. Gilbert, "Location, Location, Location: The Misprediction of Satisfaction in Housing Lotteries," *Personality and Social Psychology Bulletin* 29, no. 11 (2003): 1421–32.

9. These numbers come from the useful if overly rosy charts and statistics book by Theodore Caplow, Louis Hicks, and Ben J. Wattenberg, *The First Measured Century: An Illustrated Guide to Trends in America* (Washington, D.C.: AEI Press, 2000).

10. Gregg Easterbrook, *The Progress Paradox: How Life Gets Better While People Feel Worse* (New York: Random House: 2003), 17. As with the book above, this too suffers from a too sunny view of today's poor.

11. Ronald Inglehart and Jacques-René Rabier, "Aspirations Adapt to Situations — But Why Are the Belgians So Much Happier Than the French? A Cross-Cultural Analysis of the Subjective Quality of Life," in *Research on the Quality of Life*, ed. Frank M. Andrews (Ann Arbor, MI: Institute for Social Research, 1986), 46.

12. Lane, *Loss of Happiness*, 62.

13. See Pleij, *Dreaming of Cockaigne*, 387.

14. Christian Schneller, *Märchen und Sagen aus Wälschtirol: Ein Beitrag zur deutschen Sagenkunde* (Innsbruck: Wagner'schen Universitäts-Buchhandlung, 1867), 9–10.

15. Perrault added a coda that specifically warned girls not to be seduced by the wolf—indeed, to be most wary, not of the violent attacker, but of the nicest, sweetest-talking, and seemingly gentlest of suitors. Charles Perrault, *Histoires ou contes du temps passé, avec des moralités: Contes de ma mère l'Oye* (Paris, 1697), as cited in Andrew Lang, *The Blue Fairy Book* (London, n.d., ca. 1889), 51–53.

16. Aesop gives us another fairy tale in which a wolf, passing by a cottage, overhears a nurse tell a baby to stop crying or she will give him to the wolves. The wolf does not realize that this is an idle threat and settles in for the feast. (He is later killed.) This is not baby care at its best, but in an illustration of the story we see that a broom lies in the foreground, abandoned midsweep because of the baby's wailing. This nurse is angry enough to be dangerous. The threat of the wolf could be a symbol for a lot of violence, fear, and desire. Marina Warner, *From the Beast to the Blonde: On Fairy Tales and Their Tellers* (New York: Noonday, 1994), 296. The illustration is an etching by F. Barlow, entitled "The Nurse and the Wolf," from *Aesop's Fables* (1723).

CHAPTER 10: SHOPPING IN ABUNDANCE

1. Cited in David Brownstone, *Island of Hope, Island of Tears* (New York: Rawson, 1979), 17.

2. For more on the Uneeda story and much else of interest, see Harvey Levenstein, *Revolution at the Table: The Transformation of the American Diet* (Berkeley and Los Angeles: Univ. of California Press, 2003), 35–36.

3. As cited in Jackson Lears, *Fables of Abundance: A Cultural History of Advertising in America* (New York: Basic Books, 1944), 329.

4. As cited in Lears, *Fables of Abundance*, 244.

5. *Online Etymology Dictionary* (http://www.etymonline.com/), November 2005.

6. I borrow these two ads (for Max Factor and Pepsi) from Catherine Orenstein's *Little Red Riding Hood Uncloaked: Sex, Morality, and the Evolution of a Fairy Tale* (New York: Basic Books, 2002). Images of the ads are reproduced on pages 169 and 124–25 of that book.

7. Frank O'Hara, "Meditations in an Emergency," in *Meditations in an Emergency* (New York: Grove, 1957), 38.

CHAPTER 11: WHAT MONEY STOLE

1. Howard Mumford Jones, *The Pursuit of Happiness* (Ithaca, NY: Cornell Univ. Press, 1953), 12–14. This is a great little book.

2. Diocletian brought cabbage knowledge from Rome to Croatia. These were open-leaved plants. Hildegard of Bingen lets us know that by the twelfth century, head cabbages were in Europe.

3. It's easy to like any doctrine so antithetical to evangelism that it cannot manage its own defense.

4. Franklin to Dr. Forthergill, 1764, in *The Writings of Benjamin Franklin*, ed. Albert H. Smyth (New York: Macmillan, 1905–1907), 4:221.

5. *Writings of George Washington*, ed. John C. Fitzpatrick (Washington, DC: GPO, Bicentennial Editions, 1931), 35, 432.

6. Jefferson to Edward Rutledge, 27 December 1796. Letter 202 in *Memoir, Correspondence, and Miscellanies, from the Papers of Thomas Jefferson*, ed. Thomas Randolph, vol. 3 (Boston: Gray and Bowen, 1930). The full text is available on the Gutenberg Project's Web site: http://www.gutenberg.org/dirs/1/6/7/8/16781/16781-h/16781-h.htm.

7. Jefferson to Thaddeus Kosciusko, 26 February 1810. Letter 88 in *Papers of Thomas Jefferson*, vol. 4.

8. Alexis de Tocqueville, *Democracy in America*, trans. George Lawrence (Garden City, NY: Doubleday, 1969), 513. The title of this essay is "Of the Uses Which the Americans Make of Public Associations"; it is found in vol. 2, sec. 2. The full text is available on the University of Virginia Web site: http://xroads.virginia.edu/~HYPER/DETOC/toc_indx.html.

9. Tocqueville, *Democracy in America*, 517.

10. Theda Skocpol, "How Americans Became Civic," in *Civic Engagement in American Democracy*, ed. Theda Skocpol and Morris P. Fiorina (Washington, DC: Brookings Institution Press, 1999).

11. She says that my grandfather, Irving, was not a very social man but still spent a lot of time at the B'nai Brith. He brought the bagels. He also carted away the stale ones and fed them (gratis) to the ponies at the nickel pony rides that then existed on East 53rd Street.

12. Germans were the largest group of immigrants for a while and they were particularly associational. Still, they weren't alone in this. By 1910, two-thirds of all Poles in America reportedly belonged to at least one of the seven thousand Polish associations. The numbers for Jews, Slovaks, and Croats were similar. Steven Diner, *A Very Different Age: Americans of the Progressive Era* (New York: Hill and Wang, 1998), 9. Charmingly, in senior communities that hold such associations—Jewish retirement communities for example—there are similar friendly clubs, now gathering people from Chicago or Los Angeles.

CHAPTER 12: HOW WE BUY BACK WHAT MONEY STOLE

1. Robert D. Putnam, *Bowling Alone: The Collapse and Revival of American Community* (New York: Simon and Schuster, 2000), 367. Juliet B. Schor's *Overworked American* comes at the problem of community from the other direction: wealth and technology were supposed to lead to more leisure, but they have so far done quite the opposite in the United States. Juliet B. Schor, *The Overworked American: The Unexpected Decline of Leisure* (Basic Books, 1992).

2. Consider one reporter's description of the feeling that, with the win, something had been lost: "But there was a pain in it that hearty New Englanders began to embrace. It was part of the heritage of this notoriously difficult place to live, a place where the streets weren't wide enough and summer wasn't long enough. It was the place where spring never came and winter never left and the Sox never won. You needed a strong constitution to live in New England and a stronger one to be a Red Sox fan. Until Oct. 27, 2004. Then it was suddenly all washed away." Ron Borges, "After Sweep, Red Sox Fans Ask: Now What?: Curse of Bambino Ends, Leaving New Englanders with One Worry," *NBCSports.com*, October 28, 2004, http://www.msnbc.msn.com/6350372.

3. Edith Wharton, *The Custom of the Country* (New York: Bantam Classics, 1991), 115.

4. M. E Cain, C. M. Smith, and M. T. Bardo, "The Effect of Novelty on Amphetamine Self-Administration in Rats Classified as High and Low Responders," *Psychopharmacology* 176 (2004): 129–38; M. T. Bardo and L. P. Dwoskin, "Biological Connection Between Drug and Novelty Seeking Motivational Systems," in *Motivational Factors in the Etiology of Drug Abuse*, ed. R. A. Bevins and M. T. Bardo (Lincoln: Univ. of Nebraska Press: 2004), 127–58.

5. Most neuromarketing researchers today use functional magnetic resonance imaging (fMRI) machines that generate images of the brain as it responds to stimuli. But since the machines weigh thirty-two tons, studies had been confined to the lab. Neuroco uses electroencephalography, or EEG—a lighter and cheaper technology that allows you to follow consumers to shopping malls.

6. Cited in Tara Parker-Pope, "This Is Your Brain at the Mall: Why Shopping Makes You Feel So Good," *Wall Street Journal*, December 6, 2005, Personal Journal, sec. D, 1. Another article on Lewis is, amusingly, called "This Is Your Brain on Advertising"; this one is by Thomas Mucha and appeared in the August 4, 2005, edition of the online journal *Business 2.0* (http://money.cnn.com/magazines/business2/). The coincidence of titles demonstrates how surprised we are that experiences cause chemical "highs."

7. Sinclair Lewis, *Babbitt* (New York: Signet Classic, 1961), 81.

8. Rilke to Witold von Hulewicz, 1925, in *Letters of Rainer Maria Rilke, 1910–1926*, trans. Jane Bannard Greene and M. D. Herter Norton (New York: Norton, 1947), 2:374–75.

BODIES

1. National Safety Council, "Deaths and Injuries in the Workplace, Home and Community, and on Roads and Highways," in *Report on Injuries in America, 2003*. Data from *Injury Facts*, 2004 ed., NSC.org.

2. Epicurus, "Letter to Monoeceus." There are various translations on the Web and in print; see, for example, http://epicurus.info/etexts/Lives.html#XXVII.

3. Epicurus, "Principle Maxims." The full text is available on the Epicurus Web site: http://epicurus.info/etexts/Lives.html#XXXI. Seneca Epistle XIX, 10.

4. Epicurus, "Letter to Monoeceus."

CHAPTER 13: EATING

1. As cited in Ronald L. Numbers, *Prophetess of Health: A Study of Ellen G. White* (New York: Harper and Row, 1976), 18.

2. As cited in Levenstein, *Revolution at the Table*, 22.

3. Clarence S. Darrow, *Farmington* (Chicago: McClurg, 1904), See http://www.law.umkc.edu/faculty/projects/ftrials/dar_farm.htm for a searchable excerpt of this text.

4. Elder Roswell Cottrell, "The Health Reform," cited in Numbers, *Prophetess of Health*, 86.

5. Horace Fletcher, *Fletcherism: What It Is; or, How I Became Young at Sixty* (New York: Frederick Stokes, 1913). See also his *Happiness As Found in Forethought Minus Fearthought* (New York: Frederick Stokes, 1910).

6. Francis W. Crowninshield, *Manners for the Metropolis* (New York: Appleton, 1909), 40.

7. Horace Fletcher, *The New Glutton or Epicure* (New York: Frederick Stokes, 1906), 11–12.

8. *Harvard Crimson*, November 1905, as cited in Levenstein, *Revolution at the Table*, 92.

9. See, for example, J. B. Huber, "Do We Eat Too Much?" *Scientific American* 97 (September 1909): 167, 257–58.

10. Upton Sinclair, *The Autobiography of Upton Sinclair* (London: W. H. Allen, 1963), 120.

11. Ellis Parker Butler, "The Bone-Crackers: A Dietetic Comedy in One Act," *Munsey's Magazine*, May 1911, 203–8.

12. Upton Sinclair, *The Profits of Religion* (1918).

13. Levenstein, *Revolution at the Table*, 119.

14. Quoted in Gina Kolata, "Which of These Foods Will Stop Cancer? (Not So Fast)," *New York Times*, Science Times, September 27, 2005. Kolata is a terrific science writer.

15. Carla H. van Gils et al., "Consumption of Vegetables and Fruits and Risk of Breast Cancer," *Journal of the American Medical Association* 293 (January 12, 2005): 183–93. The *JAMA* summary of the article reads: "The intake of vegetables and fruits has been thought to protect against breast cancer. Most of the evidence comes from case-control studies, but a recent pooled analysis of the relatively few published cohort studies suggests no significantly reduced breast cancer risk is associated with vegetable and fruit consumption." The *New York Times*'s brief coverage of the news, a story called, "Eat Your Spinach Anyway," advises the following: "The study, the largest such to date, began in 1992 with more than 500,000 European participants. But there are plenty of good reasons to eat fruits and vegetables, researchers noted; they can keep the heart healthier and the weight down." This plays down the study's amazing news, as if the command had obvious authority, whatever the facts. *New York Times*, National Briefing, January 12, 2005.

16. Kolata, "Which of These Foods."

17. Kolata, "Which of These Foods."

18. Ross L. Prentice et al., "Low-Fat Dietary Pattern and Risk of Invasive Breast Cancer: The Women's Health Initiative Randomized Controlled Dietary Modification Trial," *Journal of the American Medical Association* 295 (February 8, 2006): 629–42; Shirley A. A. Beresford et al., "Low-Fat Dietary Pattern and Risk of Colorectal Cancer," *Journal of the American Medical Association* 295 (February 8, 2006): 643–54; Barbara V. Howard et al., "Low-Fat Dietary Pattern and Risk of Cardiovascular Disease," *Journal of the American Medical Association* 295 (February 8, 2006): 655–66.

19. Gina Kolata, "Low-Fat Diet Does Not Cut Health Risks, Study Finds," *New York Times*, February 8, 2006, 1.

20. Kolata, "Low-Fat Diet," 1.

CHAPTER 14: EXERCISE

1. Steven Blair, "Physical Fitness and All-Cause Mortality," *Journal of the American Medical Association* 26, no. 17 (November 3, 1989): 2395–401.

2. JoAnn E. Manson et al., "A Prospective Study of Walking as Compared with Vigorous Exercise in the Prevention of Coronary Heart Disease in Woman," *New England Journal of Medicine* 341 (1999): 650–58, as cited in Gina Kolata, *Ultimate Fitness: The Quest for Truth About Exercise and Health* (New York: Farrar Straus & Giroux, 2003), 63–64.

3. Elliott J. Gorn, "Sports Through the Nineteenth Century," in *The New American Sport History: Recent Approaches and Perspectives*, ed. S. W. Pope (Champaign: Univ. of Illinois Press, 1997), 49. See also Robert Higgs, *God in the Stadium: Sports and Religion in America* (Lexington: Univ. Press of Kentucky, 1995).

4. *Newport Croquet Club Handbook* (Newport, RI, 1865).

5. G. Stanley Hall, *Adolescence* (New York: Appleton, 1904), 2:636, 633. For a discussion of these themes, see Cynthia Eagle Russet, *Sexual Science: The Victorian Construction of Womanhood* (Cambridge, MA: Harvard Univ. Press, 1989).

6. The phrase is from nineteenth-century theorist George Beard. As cited in Russett, *Sexual Science*, 118.

7. George Beard, *American Nervousness* (New York: Putnam's Sons, 1881), 98–99.

8. From "Principles of the American Turners." The "Principles" appears on the Web sites of the American Turners and in their publications.

9. Kolata, *Ultimate Fitness*, 47.

10. Kolata, *Ultimate Fitness*, 48.

11. Jane E. Brody, "Fit Is One Thing; Obsessive Exercise Is Another," Personal Health, *New York Times*, August 9, 2005.

12. Kolata, *Ultimate Fitness*, 262–67.

CHAPTER 15: SEX

1. Onan had a religious duty to impregnate his brother's widow; he spilled his seed to avoid it.

2. Simon André Tissot, *A Treatise on the Diseases Produced by Onanism* (New York, 1832); facsimile reprint edition in *The Secret Vice Exposed! Some Arguments Against Masturbation*, ed. C. Rosenberg and C. Smith-Rosenberg (New York: Arno Press, 1974), as cited in John Money, *The Destroying Angel* (Buffalo, NY: Prometheus Books, 1985), 53.

3. John Harvey Kellogg, *Plain Facts for Old and Young* (Burlington, IA: F. Segner, 1888), 295.

4. Money, *Destroying Angel*, 13. Money adds that many in his generation avoided medical exams all their lives because they feared the doctor would read their vice in their symptoms.

5. John Harvey Kellogg, *The Art of Massage: A Practical Manual for the Nurse, the Student and the Practitioner* (Battle Creek, MI: Modern Medicine Publishing Co., 1895; repr. 1909, 1919, 1923; Mokelumne Hill, CA: Health Research, 1975), 156.

6. Kellogg, *Art of Massage*, 139.

7. John Harvey Kellogg, *Rules for Right Living* (Battle Creek, MI: Health Extension Department, 1947), 11.

8. American Medical Association Committee on Human Sexuality, *Human Sexuality* (Chicago: American Medical Association, 1972), 40.

9. Diderot argued that monogamy and abstinence were both pointless superstitions that went against nature and pleasure. For a great selection of sources, see

Michel Feher, ed., *The Libertine Reader: Eroticism and Enlightenment in Eighteenth-Century France* (New York: Zone, 1997).

10. Marie Stopes, *Married Love: A New Contribution to the Solution of Sex Difficulties* (New York: Eugenics Publishing, 1918).

11. Stopes, *Married Love*, 81–82.

12. Stopes, *Married Love*, 61 (emphasis hers).

13. Stopes, *Married Love*, 107.

14. Stopes, *Married Love*, 108.

15. For the complete results of the study, see Edward O. Laumann, John H. Gagnon, Robert T. Michael, and Stuart Michaels, *The Social Organization of Sexuality: Sexual Practices in the United States* (Chicago: Univ. of Chicago Press, 1994).

16. Laumann et al., *Social Organization of Sexuality*, 86–87. See also Robert T. Michael, John H. Gagnon, Edward O. Laumann, and Gina Kolata, *Sex in America: A Definitive Study* (New York: Warner Books, 1995).

17. Laumann et al., *Social Organization of Sexuality*, 104.

18. Laumann et al., *Social Organization of Sexuality*, 106.

CHAPTER 16: TREATMENTS

1. In the common personifications of each of the humors, Sanguinicus was often female, strong, and beautiful. Flegmaticus was calm but dull. Melancholicus was sentimental and indolent. Cholericus, of the yellow bile, was energetic and angry.

2. These actions were often also diagnostic (you bleed someone so you can examine their blood), or they might be taken to speed the illness through.

3. Montaigne, "Apology for Raymond Sebond," in *Complete Essays*, 362.

4. If you saw the HBO series *Rome*, you saw people receiving the treatment and may have assumed they were being shaved.

5. Numbers, *Prophetess of Health*, 48.

6. Athanasius, *The Festal Letters of Athanasius*, ed. William Cureton (London: James Madden, 1848), http://www.tertullian.org/fathers/cureton_festal_intro.htm.

7. You have seen the building. Rouen's cathedral is the one that impressionist Claude Monet painted over and over, in various revelations of sunlight.

8. See Glenn Uminowicz, "Recreation in a Christian America," in *Hard At Play: Leisure in America, 1840–1940*, ed. Kathryn Grover (Amherst: Univ. of Massachusetts Press, 1992), 8–38.

9. *Baedeker's Guide to the United States* (New York: Charles Scribner's Sons, 1893), 222.

10. Aaron E. Ballard, "Bathing," in *The Annual Report of the President of the Ocean Grove Camp-Meeting Association* (1875), 59. As cited in Uminowicz, "Recreation," 24.

11. First printed in the *Asbury Park Journal* and later in an article in the *Long Branch (NJ) Record* in July 1889.

12. Stephen Crane, "On the Boardwalk: Aug. 14, 1892," in *The Works of Stephen Crane*, ed. Fredson Bowers (Charlottesville: Univ. Press of Virginia, 1973), 8:515–16.

13. *Asbury Park Evening Press*, June 11, 1936.

14. *Asbury Park Press*, quoted in the *Long Branch (NJ) Daily Record*, March 24, 1905, as cited in Uminowicz, "Recreation," 30–31.

15. *Asbury Park Evening Press*, June 11, 1936.

CHAPTER 17: GREEK FESTIVAL

1. Walter Burkert, *Greek Religion* (Cambridge, MA: Harvard Univ. Press, 1985), 110.

2. Across eight hundred years, *college* went from meaning "military service" to meaning "school."

3. Burkert, *Greek Religion*, 241.

4. "Homeric Hymn to Demeter," in *Hesiod: The Homeric Hymns and Homerica*, trans. Hugh G. Evelyn-White (Cambridge, MA: Loeb Classical Library, 1914). For recent insights on the matter, see Rachel Zucker, *Eating in the Underworld* (Middletown, CT: Wesleyan, 2003); and Louise Gluck, *Averno* (New York: FSG, 2006). Both are books of poetry.

5. Douglas M. MacDowell, *Aristophanes and Athens* (Oxford: Oxford Univ. Press, 1995), 259.

6. Elaine Fantham, Helene Peet Foley, Natalie Boymel Kampen, Sarah B. Pomeroy, and H. Alan Shapiro, *Women in the Classical World* (New York: Oxford Univ. Press, 1994), 87.

7. Keith Bradley, "Images of Childhood," in *Plutarch's "Advice to the Bride and Groom" and "Consolation to His Wife,"* ed. Sarah B. Pomeroy (Oxford: Oxford Univ. Press, 1999), 184.

8. Ian Johnston translation, http://www.mala.bc.ca/~johnstoi/euripides/Bacchae_Introduction.htm.

9. William Lyman Underwood, *Wild Brother*, 1921, quoted in Marina Warner, *From the Beast to the Blonde: On Fairy Tales and Their Tellers* (New York: Noonday, 1994), 304. Skeptical? There are reports and photographs of human women nursing monkeys in Amazonia, and piglets; and I have seen a photograph of a woman nursing an orphaned bear cub.

10. Burkert, *Greek Religion*, 289.

11. Jane Ellen Harrison, *Prolegomena to the Study of Greek Religion* (Princeton, NJ: Princeton Univ. Press, 1991), 436.

12. Plutarch, *"Consolation to His Wife."*

13. Burkert, *Greek Religion*, 292.

14. Harrison, *Prolegomena*, 580.

CHAPTER 18: MEDIEVAL CARNIVAL

1. With women and slaves denied the vote, it is easy to do the math and see that over half the people were disenfranchised.

2. Mikhail Bakhtin, *Rabelais and His World*, trans. Hélène Iswalsky (Indiana: Indiana Univ. Press, 1984), 78.

3. Bakhtin, *Rabelais and His World*, 79.

4. Bakhtin, *Rabelais and His World*, 90.

5. *The Catholic Encyclopedia* (1907), s.v. "Miracle Plays and Mysteries."

6. Bakhtin, *Rabelais and His World*, 7.

7. Natalie Zemon Davis, *Society and Culture in Early Modern France* (Stanford, CA: Stanford Univ. Press, 1975), 139.

8. Zemon Davis, *Society and Culture*, 99.

9. Bakhtin, *Rabelais and His World*, 81.

10. See the essay "A Bourgeois Puts His World in Order: The City as Text," in Robert Darnton's *The Great Cat Massacre: And Other Episodes in French Cultural History* (New York: Basic Books, 1999), 107–43.

11. Zemon Davis, *Society and Culture*, 109.

12. When many men are engaged in activities that occasion their dressing up as women, it seems reasonable to guess that some of them are doing it for pleasure. Stephen Orgel's study sees theatrical cross-dressing as transvestism with all the implied sexuality and power significance. Orgel, *Impersonations: The Performance of Gender in Shakespeare's England* (Cambridge: Cambridge Univ. Press, 1996).

13. The play, *El vergonzoso en palacio* ("The Bashful Man in the Palace"), was written by Tirso de Molina, pseudonym of Gabriel Téllez, an outstanding dramatist of the golden age of Spanish literature. He was a monk and also a theologian of repute. An English translation is to be found in *The Bashful Man at Court; Don Gil of the Breaches Green; The Doubter Damned*, trans. John Browning and Firoigio Minelli (Ottawa: Dovehouse Editions, 1991).

14. There is a heated historical debate over these two interpretations. For an overview, see Chris Humphrey, *The Politics of Carnival: Festive Misrule in Medieval England* (Manchester, UK: Manchester Univ. Press, 2001).

15. Bakhtin was living in Soviet Russia, often under terrible privations, and he understood carnival as a world apart from the regular world, where one could truly feel and express oneself. This parallel world of folly, pleasure, relief, and release was just as significant in the average person's life, as was the regular world. Bakhtin saw his modern times as gray and tame in comparison to the revels of the past. For such assessments see Davis, 103, and Humphrey, 29.

16. Bakhtin, *Rabelais and His World*, 13.

17. Zemon Davis, *Society and Culture*, 139.

18. These rituals, interestingly, made it to the new world and became *shivarees*. American shivarees started as a similar critique of problem marriages, and ended up as a good-natured custom carried out for all newlyweds—people made a clatter of music under the wedding-night window and the groom paid them off with drinks. The word *skimmington* made it to the new world, too, along with its gesture of shaking a ladle to shame a woman who was too mean to her man.

19. It may seem like a lot, but bear in mind that most work was seasonal. (Farms have down times, mill rivers freeze and flood, soldiering has long breaks.)

20. Bakhtin, *Rabelais and His World*, 91.

CHAPTER 19: TODAY'S NEWS AND VIGILS

1. "The Strange Ordeal of Elizabeth Smart," *Sydney Morning Herald*, March 15, 2003. I use this distant source because most newspapers quoted only the father's first line, regarding hell.

2. Donald W. Winnicott, *Playing and Reality* (London: Basic Books, 1971), 32.

3. Sigmund Freud, *Three Essays on the Theory of Sexuality* (London: Standard, 1963), 172.

4. Martin Amis, "The Queen's Heart: In Time for Her Golden Jubilee, Two Biographies of Elizabeth II," *New Yorker*, May 20, 2002, http://www.newyorker.com/critics/books/?020520crbo_books1.

5. Percy Bysshe Shelley, "An Address to the People on the Death of the Princess Charlotte" (1817), in *Shelley's Poetry and Prose*, ed. Donald H. Reiman and Neil Fraistat (New York: W. W. Norton, 2002).

6. The three men were Jeremiah Brandreth, William Turner, and Isaac Ludlam. The crime they were accused of is referred to as the Pentrich Rising and took place in June 1817, late in the fifty-nine-year reign of George III (d. 1820).

7. Despite George III's having produced fifteen legitimate children, only his eldest son gave him a legitimate grandchild, Charlotte. When she died, all her aging bachelor uncles rushed to wedlock, the winner, at the age of fifty-nine, producing Victoria, who took the throne after a few old, childless uncles ruled briefly.

8. Stephen C. Behrendt, *Royal Mourning and Regency Culture: Elegies and Memorials of Princess Charlotte* (London: Macmillan, 1997).

9. Shelley, *Poetry and Prose*.

10. Avner Ben-Amos, "Les funérailles de Victor Hugo: Apothéose de l'évènement spectacle," in *Les lieux de mémoire*, ed. Pierre Nora, vol. 1, *La République* (Paris: Gallimard, 1984), 473–522.

11. "What Difference Does a Mourning Day Make?" *Age*, January 15, 2005.

12. "What Difference."

13. Louise Crossen, "A Nation Gently Weeps," *Age*, January 15, 2005.

14. These stories were published as a book: *Portraits: 9/11/01: The Collected "Portraits of Grief" from The New York Times* (New York: Times Books, 2002).

CHAPTER 20: WEDDINGS, SPORTS, POP CULTURE, AND PARADES

1. I especially like the Hermitage version, sometimes called *Dance II*.

2. Milan Kundera, *The Book of Laughter and Forgetting* (New York: Harper Perennial, 1999), 63.

3. Fans carry cigarette lighters in their pockets. I have seen people hold up their lighted cell phones at contemporary shows.

4. You can see thousands of pictures of the parade on the Internet.

CONCLUSION: THE TRIUMPH OF EXPERIENCE

1. Montaigne, *Complete Essays*, 403.

2. Montaigne, *Complete Essays*, 437.

Index